# Outwitting Deer

# Outwitting Deer

## Bill Adler Jr.

Lyons Press

# Contents

*From Marv Newland's film* Bambi Meets Godzilla

"Bambi Meets Godzilla" is available from:

The Rocketship Reel
PPI Entertainment Group
88 St. Francis Street
Newark, NJ 07105

    or

The Whole Toon Catalog
Facets Multi-Media
1517 W. Fullerton Ave.
Chicago, IL 60614
Phone: 800-331-6197
Fax: 773-929-5437

# Acknowledgments

No book is an island. There is always a cast of characters involved in a book's research, writing, and publication. But whenever I write the acknowledgments, I'm afraid that I've forgotten somebody, and I'm sure that is the case here. So if I've left you out, let me know and I'll remember you for *Outwitting Deer II: Bambi's Revenge*.

I want to express my appreciation for all the hard work and insight that Peggy Robin, my wife, harshest critic, and greatest friend, gave to this book. Without Peggy, you'd be reading mush. Pure and simple mush. Peggy did for this book what these deer-thwarting techniques will do for your garden: keep it alive.

Lilly Golden, The Lyons Press's ace editor, was a tremendous asset. Her advice was surpassed only by her incredible patience.

Tracy Quinn devoted countless hours and brain cells to help bring this book together. To say that she knows a lot about deer would be like saying that deer know a lot about breaking into gardens—an understatement. Her dedication and devotion to *Outwitting Deer* were incomparable.

Jessica Allen worked so hard on this book that I think the computer keyboard would run away from her if it could. Jessica was instrumental in helping put all the pieces of *Outwitting Deer* together.

Thanks, too, to Marv Newland, who created, wrote, and produced the classic film *Bambi Meets Godzilla.*

Thank you to the following people for answering questions about gardening and deer-resistant products and plants, deer-stopping techniques, facts about deer, deer they've grown to hate, thoughts about Bambi (or, more accurately, bad thoughts about Bambi), and other ideas about keeping deer out of gardens. In no particular order, I want to thank: Cliff Bingham, Jeff Chorba, Steve Graf, Diane Manning, Stuart Staniford-Chen, Ingrid Jakens, Mary Clark, Dennis Mathiasen, Lynne Mellinger, Sara Shepherd, Vaudeth Oberlander, Ed Nortons, Fred and Polly Boggs, Norwood H. Keeney Jr., David Wright, A. J. Hicks, Anita Jackson, A. M. Anz Jr., Sue Heller, Sharon Davis, Judy Pineda, Christine Reetz, S. Perry, Trisha Moore, Michael Geilich, Bernie Sayers, Martha E. Wells, Laurie Klarman, Henry Bermon, Janine Brennan, Judy Scott, Mary McGrory, and James Swann.

# Introduction

*This is how to outwit deer. Build a brick wall, tall and strong, so they can't jump over. And that's how to outwit deer.*
—Karen Adler, age 6

That's one solution, of course. But while it's a great testimonial to the imagination of six-year-olds, this isn't the most practical solution for most people. Karen also suggested the following: "You should plant no things and only buy food. Plant no flowers or anything, so the deer won't come."

It all started in 1989, the year after my book *Outwitting Squirrels,* was published. I received a call from my brother-in-law, who lives just a few blocks away. I was in the process of writing the sequel, *Outwitting Critters,* and had become the animal expert in town. "Bill, there's some strange animal in our yard," he said. Richard lives about a quarter mile from the National Zoo in Washington, D.C., home to all sorts of exotic creatures. "I think this is an animal that's escaped from the zoo. Who should I call?" he asked.

I wondered how an animal *could* escape from the zoo. It would be difficult for any animal, especially a large one, to slip out unnoticed, but stranger things have happened. This is Washington,

3

D.C., after all. (Take, for instance, the incident involving the deadly snake that was stolen from the zoo. The snake bit the thief on the bus ride home. Bad career move if that kid wanted to become a zookeeper.) I asked Richard what the animal looked like. "It's large—about the size of a donkey. And it has horns. Do you think this is some kind of gazelle?"

I may know a thing or two about outwitting squirrels, pigeons, ants, raccoons, and skunks, but outwitting gazelles isn't on my list of talents. Still, I like a challenge. "Can you tell me more about the animal?" I asked. Richard could only offer a few more details, since it was getting dark. "I think it has about four horns on its head. It's not doing much except eating the plants in our yard."

"Could those horns be . . . antlers?" I asked Richard.

"Yes."

"Then it's a deer."

"A deer?"

"But what would a deer be doing in a city?"

What indeed?! Don't those deer know that cities are off limits to them? That they are woodland animals that nibble on berries, nuts, leaves, and other forest delights? Deer are supposed to frolic quietly in the woods, leaving soft footprints, making baby deer, not bothering anybody—and occasionally succumbing to a harsh winter, a hunter's gun, or Godzilla.

But the truth is different. Like many people, despite a childhood filled with images of Bambi and many adult years spent reading *Bambi* to my two daughters, I know better. Deer are not sweet, carefree, or affectionate. No, indeed. Do you know what deer are? Deer are very, very big squirrels. This is such an important point it bears repeating: *Deer are very, very big squirrels.* All you have to do is think about it for a moment, and you'll know I'm right. Like squirrels, deer invade our yards, eat food intended for others, sometimes run into our home, leave their droppings behind, and try so hard to look cute. The only differences between deer and squirrels are that deer outweigh squirrels by a few hundred pounds and that they don't particularly like sunflower seeds. But that could change. The following chart demonstrates that deer and squirrels are either the same species or are quickly evolving into the same species.

## Similarities between Deer and Squirrels

| Squirrels | Deer |
| --- | --- |
| Come onto our property uninvited | Come onto our property uninvited |
| Pretend to have cute, "aw shucks" faces | Pretend to have cute, "aw shucks" faces |
| Eat things that are not meant for them and that are very expensive | Eat things that are not meant for them and that are very expensive |
| Eat sunflower seeds | Eat flowers in the sun |
| Come in boring colors, mostly gray, black, and white | Come in boring colors, mostly brown and white |
| Sometimes sneak into our houses, then can't figure their way out | Sometimes sneak into our houses, then can't figure their way out |
| Carry diseases—if you're bitten | Carry diseases—ticks do the biting |
| Run away when children try to play with them | Run away when children try to play with them |
| Like to pose for pictures, as if that compensates for all the trouble and expense they cause | Like to pose for pictures, as if that compensates for all the trouble and expense they cause |
| Frequently found flattened underneath cars | Frequently found as hood ornaments on cars |

But despite these similarities, there's one major difference between deer and squirrels: The techniques that you use to outwit squirrels just won't work for deer. No matter how much you wish it were otherwise. You can't grow your tulips beneath a baffle. You can't string your carrot plants along a wire and between LP record albums. You can't grease your tomatoes with Vaseline.

A lot of people like deer. To tell the truth, I used to be one of them. Why did I like deer? Because I'm from New York City, and just about any wildlife—pigeons and rats excluded—is attractive and interesting. As a New Yorker I once believed all those myths about Bambi. I saw the movie. But the truth is very different. They invade our yards, eating our precious plants, shrubs, fruits, and vegetables. That's why they need to be outwitted. They ruin our pretty paths with their scat and tracks. They leave their ticks behind to infect us and our pets with Lyme disease and who knows what else. They crash into our cars on the highways, and, well, generally make a nuisance of themselves. The truth is, deer are a scourge. Never fear, however: Deer *can* be outwitted, and I'm going to show you exactly how to do it.

## Interviews with Bambi and His Alter Ego, Bruno

Over the years Bambi has acquired a reputation as a gentle, harmless creature, frolicking through the woods. And that's true. Bambi is a harmless creature—but more than that, Bambi is adorable, affectionate, pretty, and pleasant to be around. But it's also true that Bambi doesn't exist. The truth is that if all deer were like Bambi, there would be no need for *Outwitting Deer.*

Deer are more like the antithesis of Bambi: They are Bruno. Bruno is a deer that prefers gardens to the ample foliage of the deep woods. Bruno is a deer that, when he finds out about a garden with great lettuce or tulips, tells all the other deer in the woods immediately. Bruno is a deer that sports the deer tick and that leaves deer droppings around to let you know exactly which plants he's been eating (as if you didn't know already). Bruno (and, in all fairness, Brunhilda) is a deer that will wander into your house if you leave the door open, and then, while trying to escape, destroy all your fine collectibles from the Home Shopping Network.

Bambi is like Casper the Friendly Ghost. And just as real. Everyone knows that there's no such thing as a friendly ghost: Ghosts are out there to frighten us, to do mayhem, and mostly to drive us away so that they can take possession of our houses and our property. The same is generally true for deer.

Bruno is the evil deer, the typical deer. Bambi is the fantasy deer. And although Bambi is no more real than a few brush strokes on a film cell, Bambi is the deer of children's dreams. To the extent that people believe there is a little Bambi in every deer (however untrue), that belief is a reality we must contend with. It is my contention that it is this deep-rooted, perhaps subconscious, notion that there's a little Bambi in every deer that prevents many of us from using flame throwers or small nuclear weapons in our backyards. We also don't want to destroy the very gardens we are trying to preserve just to get rid of these deer, right?

So now let's peer into the minds of Bambi, our fictitious, friendly deer, and Bruno, the *real* deer. I have asked each the same questions—more or less. Judge for yourself which deer you believe and trust.

| INTERVIEW WITH BAMBI | INTERVIEW WITH BRUNO |
| --- | --- |
| *Why are you so friendly?* | *Why do you invade gardens?* |
| All deer are God's creatures, and my function in nature is to frolic and make little children smile. | Because I can. And there's nothing you can do to stop me. |
| *I've heard about the deer tick, a parasite that can make people sick. What are your comments about that?* | *I've heard about the deer tick, a parasite that can make people sick. What are your comments about that?* |
| Oh my. I didn't know. Perhaps you shouldn't pet me. | That's another reason you should stay away from me. Here's the deal—you keep your distance while I take your tulips and turnips, and you won't get sick. |
| *What's your favorite activity?* | *What's your favorite activity?* |
| Being myself. I like playing in woodland glades and taking care of my fawns. | I like pole vaulting over high fences. |

BAMBI *(Continued)*

*How do you feel about hunting?*

It's so sad. I just wish people wouldn't do that. I mean, what have we ever done to you?

*Who are your best friends?*

Thumper and Flower.

*Who is your hero?*

Rudolph the Red-Nosed Reindeer.

*What do you like most about being a deer?*

Spending my time in the woodland glades, admiring nature's handiwork.

*Do you have any tips for protecting gardens from marauding deer?*

Most of the time, a posted sign will be enough. We're really polite.

BRUNO *(Continued)*

*How do you feel about hunting?*

It's okay. Really. Hunting takes care of the stupider deer, leaving the ones that are really good at breaking into gardens, like me.

*Who are your best friends?*

The squirrel next door, who breaks into bird feeders so successfully. We compare notes all the time.

*Who is your hero?*

The gopher from the movie *Caddyshack.*

*What do you like most about being a deer?*

I don't really like being a deer. What I'd really like to be is a grizzly bear. Then I'd get respect.

*Do you have any tips for protecting gardens from marauding deer?*

Do I have to answer that question?

*Not if you don't want to. But you probably know more about this than anybody.*

Well, poisoning all your veggies would help, but then I guess you wouldn't be able to make much use of them, either. How about an indoor greenhouse?

Outwitting deer is not something you can do overnight. You can't just buy an Acme Deer Outwitting Gizmo and expect that deer are going to run away scared, their white tails (if you have white-tailed deer) waving bye-bye as they frolic away down the path. You have to be willing to spend a lot of time at it—or abandon your garden. It's this simple: Deer have nothing to do all day long but figure out ways to break into your garden. They don't watch television, don't pay taxes, and don't have to take their kids to school. They have one directive in life and that is invading your garden.

But you can use proven techniques, including using plants that deer don't like, specialized fencing, chemical deer repellents, high-pressure water, big dogs, deer distractions, noisemaking devices (and I'm not talking about screaming at the deer), robots, scary lights, and more. You have to look at the world from the deer's perspective. You have to think like a deer, to look at your garden from the deer's point of view. To do this you need to understand how deer behave, eat, breed, and live. You have to wander around your yard as if you were a deer. Never mind if your neighbors think you're a bit crazy. This is war.

—Bill Adler Jr.
www.outwittingdeer.com
www.adlerbooks.com

# The Awful Truth about Deer

*I ascended to the top of the cutt bluff this morning, from whence I had a most delightfull view of the country, the whole of which except the vally formed by the Missouri is void of timber or underbrush, exposing to the first glance of the spectator immense herds of Buffaloe, Elk, Deer, & Antelopes feeding in one common and boundless pasture. . . .*

—Meriwether Lewis, *The Journals of the Lewis & Clark Expedition*, 1804–06

D eer are everywhere. They are in all forty-eight contiguous states in America, eight Canadian provinces, Mexico and Central America, northern South America, and Asia. We're surrounded. Moving to Europe isn't the solution either. Deer are a big problem in Scotland, for instance, where the proceeds from the national lottery go toward building deer fencing.

Deer are in our national parks, on our highways, in our zoos (Why?—it's not as if we actually need to go to the zoo to see deer), sometimes in our houses, and, most important, in our gardens and backyards. If deer aren't a problem in your area, you have to deal with deer relatives: Aunt Reindeer, Uncle Moose, Mama Llama, Grandma Goat, Grandpa Mule Deer, Brother Elk, Cousin Caribou; the list goes on. They all like gardens.

We can't escape deer, mostly because we now live in their habitats. Bambi's great-great-great-ancestors grazed, mated, and raised their families in the same spot you are raising yours. This was once *their* land. Since we can't avoid them, we're tired of feeding them, and we have no desire to shelter them, it's time to learn to outwit them. But first some insider information.

The more you know about deer, the better your chances of outwitting them will be. Knowledge is power—to some degree, anyway. The information I've provided should be invaluable in your efforts to deter deer. But pure knowledge can go only so far. Just as a medical student first learns by reading, then by watching, and finally by doing, so must we deer outwitters. After you read about deer, it's very important to watch them. Observe where they come from; that is, what route they take into your garden. Where do they look when they're eating your prized roses? Do they also take a drink in your pool or pond? Do they seem to be a little nervous about the neighbor's dog?

I highly recommend walking around your property and garden and looking at the work from the deer's perspective. Get down on all fours and crawl around; try to see things as they do. No, you don't have to don antlers. Only when you see your precious plants through a deer's eyes will you have a chance of outwitting the critters.

## History of Deer

Thousands of years ago people used deer the same ways that sheep, cattle, and goats are used today. People ate their meat, wore their skin as clothing, and used their hides for instruments and shelter. Various deer body parts served medicinal as well as magical purposes. And, of course, deer pulled sleds. Some people, such as the Lapps in the far northern corners of Europe, actually domesticated the deer, growing them for milk and meat. Over time, however, deer proved far less popular as a domesticated animal than the species we are familiar with today. Lucky us; after all, can you imagine a pet deer? Just picture the confrontations between neighbors: "If I ever see your unleashed deer eating our turnips again, I'm going to call the cops!"

But deer never became unpopular. Even as goats, sheep, and cattle became the standard agricultural animals, deer were still hunted. Curiously, though, venison—in other words, deer meat— became a delicacy rather than common fare despite large populations of deer.

Deer have been around for quite a long time. Biologists guess that the first deer originated approximately twenty million years ago during the end of the Miocene era. (Had deer evolved another 180 million years earlier, they would have coexisted with the dinosaurs and that would have prevented our problem.) Early deer spread throughout the forests of the northern temperate regions of the world. Other hoofed animals that evolved at roughly the same time, such as giraffes, tended to cluster in the warmer tropical regions.

## That's a Lot of Deer

It seems as if some people just can't get enough of deer: The *New Straits Times Press* in Malaysia reported in March 1997 on the opening of a deer park in Hulu Semenyih. Visitors are able to sit at picnic tables and watch about two thousand deer gallivanting beneath trees and shrubs. Officials expected the park to be a bustling attraction because it will allow city dwellers to see how deer live.

While it is impossible to determine precisely how many deer roamed North America centuries ago, many experts agree that there were on the order of twenty million white-tailed deer around the mid-1600s. By the end of the 1800s, only an estimated five hundred thousand were left, the result of massive deforestation and hunting by settlers. For those two hundred or so years there were no limits on the number of deer a hunter could kill, and there were no hunting seasons—deer were hunted year-round.

Around the early 1900s hunters began to notice this decimation

antlers; instead, the males had sharp, small tusks that they used to defend themselves. Some modern deer in China still sport tusks and no antlers. Just imagine if modern white-tailed deer still had tusks. That would make trying to outwit them a whole lot more interesting, and deer would have a lot more bargaining power when it came to negotiating a settlement about our gardens.

In Europe the roe deer, red deer, and fallow deer are the most common species. Great Britain has six types of deer: Chinese water deer, fallow deer, red deer, roe deer, sika deer, and muntjac.

In the United States, Canada, and Alaska, you can find the moose. The elk can be found in northern New England. Canada has vast herds of caribou that live on the frozen tundra. In the United States the most abundant deer species are the whitetail (found in the East), the blacktail (found along the Pacific Coast), and the mule deer (found in the West). The Rockies have a mix of whitetails and mule deer.

## Where and When to Send Deer Mail

Most species of deer live in the woods, although some species live in swamps, marshes, or grassy areas. Of course, if you let them, most species of deer would be just as happy to live in your house or garden.

We can't blame the deer invasion entirely on them. Contrary to popular belief, deer prefer the edges of woods and forests, called edge habitats, rather than the deep interior.

Developers, in an effort to give a country feel to housing complexes, often situate them in or near forests, thereby creating new edges for deer. And as urbanization continues, humans move closer and closer to deer territory. With our large lawns, tasty flowers, and budding trees, deer love us. Deer travel along the banks of man-made waterways to infiltrate cities, where they find cemeteries, golf courses, and parks as inviting to them as natural meadows.

Deer are territorial animals—like people. Just as we have our favorite spots in our house where we go to enjoy a midnight snack, deer have their favorite spots to graze, and once they've found these spots, they establish feeding patterns. And once our gardens have become these spots . . . well, you get the point.

Furthermore, deer require large ranges. Mule deer have a home range of between ninety and six hundred acres, depending on their sex (males roam farther than females). Black-tailed does live in an area from one to two hundred acres in circumference. Territoriality affects the size of the deer population. When a population reaches a certain size, it simply cannot support any more individual deer, and the newborns and their mothers are forced out into their own new territory: *your* garden.

Deer are very adaptable. They have to be. The white-tailed deer's environment, as well as the environment of most deer, changes seasonally, from periods lush with leaves, grasses, and fruit—usually in the summer—to periods where food and water are hard to find—generally in the winter. A black-tailed deer, for example, will eat not only the abundant leaves in the summer but various mushrooms and herbs as well. In the fall there's very little new growth, so deer eat fallen acorns and wild nuts.

Winter, of course, is the hardest time of the year to find food. During the winter you will see a lot of deer on their hind legs trying to get to the few leaves that remain on the trees. Along with the season's snow and ice, you'll see deer hoofprints and scat (or deer droppings). Winter is the worst time for gardeners, and the best time for deer to become garden infiltrators. Many deer starve each winter, unless they can find other food, such as what might be in *your* yard. Finding food in the winter is an imperative. So don't assume that you can neglect your garden in the winter, because the deer will find whatever's left over or planted. Once they get to know your garden in the winter, they'll be there all year long. So outwitting deer starts in the winter.

Eventually winter ends, and those deer that have survived find

ample food in the spring when leaves are possibly their most deli-
cious. (That doesn't mean deer will ignore your garden in favor of
the forest, so spring is no time to let down your guard.)

Despite winter's starkness, early spring may actually be the most
dangerous season for deer. Not only is food at a minimum—
because nothing has grown all winter and because the little that's
green has been eaten—but the deer's energy requirements are also
higher in the spring. They are hungrier and will therefore go farther to find food, including crossing that wide gray asphalt strip
with the yellow line to get to the goodies on the other side of the
road. Your first spring plantings must be accompanied by antideer
strategies.

## Society: Everybody Needs a Friend

In some species deer run in herds. Probably the largest herding
deer is the caribou; these animals travel in large groups twice during the year, journeying nearly a thousand miles in the winter in
search of food. How would you like to see hundreds upon hundreds of deer crossing your garden twice a year? Sure makes one or
two whitetails seem like a picnic.

Elk also live in herds, but males and females stay in separate
herds until mating season.

Other species of deer—whitetails, for example—do not live in
herds. One of the main reasons that deer travel in herds is for protection, the same reason that fish travel in schools. But another
reason is that one deer can help defend another if it is in a large
group. Whitetails sometimes travel in groups of two or three.

Even deer that do not travel in herds protect one another. The
bull moose, for instance, follows behind a cow, protecting her rear,
because most predators attack from behind. White-tailed does protect their fawns by running around in a zigzag pattern, attempting
to confuse the predator. As they zig and zag, they leap—this prevents their scent from being left in a continuous line.

*Deer in Woods During Winter (Henry Bermon)*

ple, when food is plentiful and predators are not, fawns become sexually mature quickly, often as young as age two. (Fortunately, the same thing does not apply to humans; otherwise the world's population would have increased tenfold since the creation of fast-food restaurants.)

For males weight is not as important. Male whitetails start producing sexual hormones toward the end of their first year and grow antlers at the end of their second. Although sexually capable, male deer don't breed unless they can successfully compete against other males for the right to mate. (A similar trait in humans would have turned us into a population of hockey players.)

Young males must usually find new territory of their own. This can be a difficult time for them because just as they are seeking new territory, older adult male deer are re-establishing their own territories and have plenty of experience at defending their turf. Not surprisingly, males that lose fights over territory have a shorter life expectancy. To make things worse (that is, from the male deer's point of view), males are also at higher risk of death in collisions with cars during mating season because their single-minded focus on mating all but eradicates their judgment. Bucks run into cars, cross highways, and travel long distances in search of those perfect mates. I'll resist the temptation to make analogies to human males here.

Male deer generally weigh from four to fourteen pounds at birth, female deer about eight pounds; these discrepancies depend on the species of deer, the health of the parents, and the overall environmental condition. Fawns of both sexes remain with their mothers after they finish nursing at about two years old. (They have now begun to eat leaves and grass.) Mother deer are much more willing to let their female children stick around longer after nursing, but some are also willing to let their male children stay.

Summer is the quiet season for deer, mostly because food is usually plentiful, but subtle changes take place in the deer population that will set the stage for mating season. Dominant bucks begin engaging in stare-downs and flailing their hooves to convince other bucks that confrontation would be futile. These behaviors help determine who will be the Big Buck when mating begins.

In late August the bucks' velvet (or the soft covering on their antlers) dries and testosterone begins flowing throughout their bodies, compelling them to rub against trees. This rubbing strengthens the neck and shoulder muscles and leaves a scent to announce their presence to others. Once free of their velvet, bucks are ready to engage in physical competition. Sparring matches are generally playful skirmishes between two bucks of equal size. Generally, it's the size of the opponent's antlers that causes one to shy away from the other. Naturally, there are occasions when these encounters become dangerous to both bucks; injuries—sometimes even fatalities—occur.

When a buck begins to get the urge to merge, his neck swells; however, bucks are capable of breeding before and after this swelling period (which lasts as long as the mating season—roughly two months).

By October the does' estrogen levels climb and they smell different, especially to bucks. By early November the bucks' testosterone and the does' estrogen levels have peaked, setting the stage for the "rut," or mating season.

In November bucks communicate among themselves with bleats, snorts, grunts, bellows, bawls, low whining noises, and even wheezes. Bucks spend their time rubbing, scraping, and searching for does. Their acute sense of smell comes into play. Bucks begin to lip-curl (or, as the biologists say, engage in "flehmening") to determine if a doe is entering estrus. A buck will smell a doe's urine and then lip-curl.

A doe entering estrus will accept the company of any and as many bucks as possible. Generally, once a buck finds a doe in estrus, he will stay with her for about three days, mating with her several times. Because the buck spends all his energy finding and mating as well as antagonizing male competitors, he may lose up to 25 percent of his body weight during the whole breeding process. (Some enterprising publisher will probably create *The Buck's Way to Speedy Weight Loss!*)

## They Will Survive

A deer's survival skills are based on two things: the ability to detect predators and the ability to escape them. The deer's eyesight is not as great as you might guess; scientists have determined that deer can see vague shapes at distances up to fifty or sixty meters and *may* be able to perceive certain colors at that distance. Farther than sixty meters deer have no ability to see color at all and probably are unable to distinguish stationary shapes.

This fact does not put deer at a tremendous disadvantage to predators, however, because deer have the ability to detect movement from distances up to three hundred meters, and it is this very special sense that prevent deer from turning into venison. If you get the chance to sneak up on a deer, you will discover that you can do so successfully if you slowly move toward it only when it's not looking at you. As long as the deer doesn't see you move, you're just going to look like a strangely shaped bush to it. It may, however, flee at the slight sounds you make as you approach; deer have excellent hearing as well as sense of movement. (Deer that have become familiar with humans, however, behave differently and may not necessarily run when they see you move. More about this in a later chapter.)

In the wild the average life span of a deer is sixteen years; in captivity, twenty-five.

## Take Two Aspirin and Go Back to the Woods

Like all creatures, deer are susceptible to infection. The parasites that can infect deer are many. They include intestinal parasites such as roundworms, tapeworms, flukes, brain worms, throat worms, eye worms, foot worms, bladder worms, colon worms, muscle worms, and lungworms. Deer are also afflicted with external parasites, including biting lice, sucking lice, mites, warble flies, botflies, and keds (bloodsucking flies). Some parasites affect only particular species of deer.

Most types of parasites don't kill the host because to do so is to sign their own death warrant. (Bacteria and viruses can get away with killing their hosts because they are so easily transmitted from one host to another. This also means that if deer are nearby, you should never drink from a stream or river without purifying the water first.) Parasites are not readily transmitted from one host to another. Having a parasite is no picnic for deer and can often make them susceptible to other infections, therefore increasing their mortality rate.

## Hungry?

The more you know about how deer eat and how they process food, the better equipped you will be to outwit them. You might not think of cows when you think of deer, but when it comes to digestion they're a lot alike. Both species have four-chambered stomachs and chew their cud. In deer this arrangement serves the clear evolutionary purpose of allowing them to eat quickly without chewing, spending as little time in the open as possible. The deer's digestive system provides it with another advantage for survival: Because the animal doesn't have to spend a great deal of time chewing its food (or, sometimes, chewing at all), it can eat and run. So a deer can obtain nutrition and energy quickly. If there's danger and the deer has to run away, it's running away with a meal inside.

In the safety of the forest deer regurgitate the food from their first stomach chamber to their mouth and chew a little more. Rechewing the food enables deer to break down plant cells that are surrounded by a cellulose wall. Mammals rely on microorganisms to aid in the process of digestion. In humans and other species these microorganisms live relatively far down in the digestive system; in deer the bacteria live farther up in the system, in the stomach itself.

The first two chambers of the deer's stomach contain the bacteria that digest cellulose. When food is regurgitated and rechewed, more surface area is created, giving the bacteria more opportunity to break down the cellulose. Deer will regurgitate several times until the cud is small enough to pass into the third chamber. When the "rumen"—the fancy name for the first chamber—becomes distended and enlarged, it sends a signal to the deer's brain to stop eating and automatically regurgitates the food. The second chamber of the stomach acts as a storage container where the food is further broken down. In the next two chambers fats and proteins are broken down (much as the human stomach does); then all the food passes to the small intestine, where absorption begins.

The first chamber of the deer's stomach is the largest. It permits a deer to gulp down a lot of food quickly. What this means from a practical perspective is that a deer can eat the entire contents of your garden faster than the time it takes you to run screaming in its direction. If a deer is interrupted in the middle of eating, then it will delay the digestive process until it feels safe. Once safe, the deer regurgitates the cud, which it slowly chews and then swallows once again. Food spends about one to two hours in the second stomach chamber, then twenty or so hours in the third and fourth chambers.

As everyone knows, deer can pretty swiftly wreck a garden. But they also have an adverse effect on forests. If there aren't too many deer in a particular region, the grazers may actually increase the size of the plant population: Certain foliage will grow faster, compensating for those plants the deer have eaten. There's some speculation that deer may have a plant-stimulating substance in their saliva. It is clear, though, that too many deer will have a significant adverse effect on plant life. Defoliation affects other animal species as well, both directly and indirectly—directly because it reduces the food supply, and indirectly because it affects the nutrient cycle in that particular area.

There is, however, competition among various species for food in any given range. In some parts of the country several species of deer may coexist in the same range. For example, in part of Washington State both the black-tailed deer and wapiti live in the same forest during the summer. Sometimes different deer species occupy different ecological niches; that is, they eat different vegetation, which usually happens during the summer when food is plentiful. White-tailed deer ordinarily prefer dense coniferous forests, moose like areas with dense shrubs, and wapiti like open areas with ground vegetation. In the winter, when snow covers the ground, all three species may compete for the same limited food. (That's assuming your garden isn't nearby.)

Deer have teeth well adapted to most garden foods. Fortunately, deer don't bite. Their front teeth are located only on their lower jaw. Deer have sets of back teeth on both the upper and lower jaws. The upper jaw is primarily used as a cutting board, with the lower teeth doing the cutting.

## Faster than a Speeding Bullet

While deer are not among the smarter species on earth, they are onc of the fastest. Deer can go from zero to forty-five miles per hour in a split second, which is a whole lot faster than my 1970 Volkswagen could do it. While running at a considerable speed, deer can scoot around boulders, leap logs, and dodge trees. Compare this to the average human adult, who typically walks into a wall, signpost, or bed frame at least once a week. You've got to give them credit.

Deer have four toes on each foot. The hoof is made up of the two inner toes and is covered by a thick toenail-like substance.

Their hooves allow deer to run on tiptoes, which is part of what gives them their great speed.

Deer legs, skinny but strong and muscular, are similarly built for speed. Their long necks enable them to accomplish two tasks: lift their heads high to listen for danger and lower their heads to nibble on your radishes, carrots, roses, and everything else in your garden. A deer's eyes are positioned to allow it to see in every direction (except backward) at the same time. Though their eyesight is not all that good—things often look blurry to them—they are, as I've noted, very good at noticing motion.

A deer routinely rises in five steps:

- The deer heaves onto its front knees.
- It then raises its hindquarters.
- It lifts a front leg forward.
- It extends one leg.
- Once standing, it stretches its muscles.

A startled rise has two short steps: The deer flexes its legs to catapult its body forward and then pivots off the rear feet and is instantly running. See how fast they can move?

## Antlers: Unique Deer Hats

Let's talk about antlers. Why? Why not? It's a lot less stressful to talk about antlers than it is to talk about deer eating at your garden. Besides, no matter how much you may dislike what deer do to your garden (or your living room if one gets into your house), you have to admit that deer antlers are attractive. Here are some antler facts.

A common myth is that a deer's antlers and the number of points indicate his age. That's wrong. You can't tell how old the male deer is by the number of points on his antlers. While the size of the antlers does give a general indication of age, the number of points depends a whole lot more on how healthy that individual is. So the more points, the more a deer has been enjoying fine dining—possibly at your expense. When a deer's antlers are small or thin, that deer is usually unhealthy; a male deer with a large rack generally has a nice round barrel of a body, too. The base of the antler is the only part that gets larger each year.

Another myth is that antlers are the same as horns. Horns are bony structures that actually grow out of the skull; in many species they appear in both the male and female. Antlers, however, always grow in pairs and aren't permanent. The blood circulating through the antlers' vascular system is cooled by the outside air, helping to cool the whole deer. And to win on *Jeopardy,* keep this in mind: Deer are the *only* species that have antlers. If it's wandering in your neighborhood at night and has antlers, it's a deer.

Have you ever imagined walking around with an entire set of encyclopedias on your head? Well, that's similar to what a male deer has to endure. A bull moose's antlers weigh eighty pounds and may measure six feet from tip to tip. The extinct Irish elk had antlers that were more than twelve feet from tip to tip and weighed three-hundred-plus pounds. Antlers take from four to five months to grow, and they require a lot of food in the meantime.

Deer antlers go through four stages of growth in response to seasonal changes, or fluctuations in daylight. The cycle is controlled by the endocrine system. Male deer have two bits of bone on their heads, similar to tree stumps, from which emerge the nascent antlers in the spring. These pre-antlers are soft and pliable and covered with "velvet," which is actually a thin skin containing a complex vascular network that carries minerals and proteins into the growing antlers. After about three months antlers reach their full size but are still covered with velvet. Testosterone—which bucks begin producing in August—causes a burst in antler growth. The velvet stays on the antlers until the antlers' bone has hardened— about four months after the antlers first started to grow, just before mating season. In the autumn, or the beginning of mating season or rut, the velvet is shed, which takes just a few days; to accelerate this process and to sharpen the antlers, deer rub them against trees. (Bare patches in the bark of your trees is another sign that you've got company.) While the antlers are covered with velvet they look a little fuzzy—hence the name—and once the velvet is gone they appear crisp, beautiful, and dangerous.

Why antlers? If antlers are such a good idea, then why don't people have them? (Aside from their being a setback for the hat business.) In addition to antlers' aforementioned cooling qualities, many people believe that deer use them as weapons of sorts when males joust to win the affection of females. In fact, antlers serve to protect deer from injury by locking in such a way that the deer cannot impale each other.

It is true that female deer are attracted to those males with the biggest antlers. This makes good evolutionary sense, because those male deer with big antlers are the healthiest and most likely to produce vigorous offspring.

Antlers, shmantlers, you say. They look good on my wall, you comment. I'm not scared of hard, sharp, bony things that protrude from dim-witted animals, you tell me. My retort? Be afraid, be very afraid. Soon enough, you are going to learn about the damage antlers—and deer—do, and I am not just talking about the damage to your garden.

# Menaces to Society

*There is no little enemy.*
      —French proverb

rom a distance it's easy to see why so many nature nuts love deer. They do look beautiful and harmless and innocent. And that blink: The way they blink their eyes says, "Who, me?" But Bambi has an evil side. Some of us who have seen that side call deer "rats with horns." I'd call them squirrels with horns.

Everyone who has had close-up deer experiences has at least one story of a deer encounter that wasn't of the aw-shucks-isn't-it-cute? variety. Not even our presidents were free of deer troubles. In a letter from his Mount Vernon home, President George Washington wrote:

> The gardener complains heavily of the injury which he sustained from my halfwit, halftame deer; and I do not well know what course to take with them. . . . Two methods have occurred, one or both combined, may, possibly, keep them out of the gardens and lawns; namely, to get a couple of rounds and whenever they are seen in, or near those places,

to fire at them with a shot of a small kind that would make them smart, but neither kill nor maim them. If this will not keep them at a distance, I must kill them in good earnest, as the lesser evil of the two.

## Close Encounters of the Deer Kind

Deer can be seen trotting along our city streets and munching their way through our suburban gardens. They've even been caught eating fish in a lake, sleeping by the interstate, jumping off bridges, feasting on ornamental gardens, and sleeping under decks. Because there are no predators in urban areas (excluding the automobile, that is), deer slowly invade suburbia. When they do, deer cause more problems than simply crushing grass to make their beds and ripping up the lawn with their even-toed hooves. If you think that garden problems are the worst they can dish out, you don't know what bad is. But you're about to find out.

Norwood H. Keeney Jr. wrote to tell me about his moose encounter:

> As you most likely are aware, the moose population in New Hampshire is rapidly increasing to the point where cars and their occupants may be heading to the endangered species list. Now to my experience with one adult cow moose. I have a three-quarter-acre wildlife pond not too far from a gravel road in the town of Unity, New Hampshire, which is in the Connecticut River Valley near Newport. In late spring I was removing new growth, brush, and weeds from the waterside on the earthen dam with a string trimmer. As happens with those trimmers, the string became entangled. As I stood on the top of the dam, which is road-width, I slowly and quite quietly worked on undoing the mess. On my left shoulder, I felt a nudge. As I slowly turned, not knowing what to expect, I found myself looking

into one very large brown eye of what seemed to be a very large cow moose. After eyeing each other for what seemed forever—most likely a minute or so—she ambled to the end of the dam and into this beaver pond and into the woods. I do not know who outwitted whom.

Deer can be especially troublesome during the Christmas season, as evidenced by the following two stories. Rudolph appears to really enjoy Christmastime in this tale. A. M. Anz Jr. tells me:

> Here is a true story that I think you will enjoy. I live in Hollywood Park, Texas, a small community on the northern edge of San Antonio. During the Christmas season in 1996 I was an investigator with the Hollywood Park Police Department. Many residents here decorate their yards during the Christmas season. One of our patrol officers received a call from an irate citizen who stated that vandals had stolen a long string of Christmas lights off one of his shrubs. A report was taken and several hours later another call was received, this time from a citizen about three or four blocks away. The report: He had seen a large buck with a string of Christmas lights wrapped around his antlers. Case solved.

In Westchester County, New York, there have been reports of deer eating wreaths off the front of doors. And they aren't that well-behaved in the summer, either. The *New York Times* (November 9, 1997) reported that during the summer in Briarcliff Manor, New York, a deer that was frightened by a passing car entered a house, knocked dishes out of the kitchen sink, broke glassware, overturned a dining room chair, and finally took a dip in the above-ground pool before leaving the premises.

Block Island, Rhode Island, is experiencing the same deer problems as many other northeastern communities. The state reintroduced deer—four does and a buck—to the island in 1968 at the request of its hunters. Due to mild winters and a lack of natural predators, the deer thrived, increasing in number dramatically; they are now estimated to be about a hundred strong. In the last nine years there have been a reported forty to forty-five cases of Lyme disease each year. (More on Lyme disease later.) Fences and various deer repellents have not worked for the community.

When the Olsen family in Stillwater, Minnesota, opened their new furniture store called Painted Dreams in June 1997, they didn't expect their first "customer" to be a deer. Around six in the morning a deer crashed through the plate-glass window and ransacked the store. The deer cut its leg badly jumping through the window and splattered blood all over the store. Passersby saw the deer and called police, who arrived and shot it.

Hungry deer are jeopardizing our livelihoods, disrupting our daily lives, and presenting a health risk. Ask Pamela Glen how bad the deer problem is. She described the following incident to the *Boston Globe:* While she played in her New England backyard with her five-year-old granddaughter, a hundred-pound buck crashed through the yard and kicked her in the side of the head, resulting in a trip to the hospital with a fractured cheekbone and a nose broken in three places.

Responding to a report of a break-in at a church in Claycomo, outside Kansas City, policemen encountered a startled buck jumping from pew to pew. They were forced to tranquilize and remove the deer from the church.

Kim and Carol Haramis operate Heritage Farm, a Christmas tree farm near Peninsula, Ohio. Deer were never a problem for them until about five years ago, and now they see twenty to thirty of them regularly. They have a "nuisance-kill permit" from the Ohio Department of Natural Resources, which allows them to shoot on sight. "People think deer won't eat Scotch pine," Haramis told the *Plain Dealer.* "I can tell you that I have thousands of dollars' loss in Scotch pine. Firs are like candy to deer. Fraser and Canaan fir are like ambrosia. They are eating our Douglas fir as well."

In June 1997 an airplane was forced to make an emergency landing after hitting a deer on a Smyrna, Tennessee, runway—the third such accident since 1988. The twin-engine plane struck the deer, which had been lying on the runway, and was forced to land twenty miles away.

As a pilot I was trained in evasive maneuvers to compensate for engine outs, oil fires, communications problems, faulty landing gear, even bird strikes . . . and deer. (You learn to keep your hand ready on the throttle to "go around"—take off again while you're landing—if you see a deer on the runway.) Deer-airplane collisions happen now and then, partly because deer don't expect any predators to come in from above.

You don't find many people walking into the local Pets Are Us and asking for a deer, but for one family in Glastonbury, Connecticut, a pet deer works just fine. According to the *Hartford Courant,* a baby fawn wandered onto their lawn one morning; instead of calling the local police department, Bill Shaw, an avid deer hunter, decided to keep her as a pet.

Shaw is not alone in his sympathy for deer. When a couple traveling from Oklahoma to Texas found a fawn about ten days old and his dead mother on the side of the road, they dropped him off at the local Humane Society, where the veterinarian surgically repaired his injured leg and inserted a stainless-steel pin. The young male deer's six- to twelve-month recovery period began in a local kennel, but because he became nervous around the dogs, a volunteer, Don Praeger, decided to take him to his Texas home. The deer now roams the Praegers' house and has even managed to change their twenty-two-year-old son's view on hunting. He told the *Dallas Morning News* that Bambi—the not-so-original name the Praegers picked for the fawn—had "changed my whole perspective on deer hunting." Although he had never shot one himself, he said he has now decided he will never go hunting again. When the fawn is fully healed, he will be released into the wild again.

## Fawn on Your Lawn

It is not unusual to find a fawn lying in the grass during the spring or early summer. If you find one in your yard without its mother, the best advice from experts is to leave it alone. Most likely the fawn has been left by its mother while she goes in search of food, and will be reunited with her soon. Although fawns look cute and cuddly, they average about ninety pounds and have strong muscles. Young deer can be very high strung and will use their strength to defend themselves when they believe they are in danger. Also, fawns learn butting as a means of practicing to be an adult; they may confuse a human with another deer and try to practice on you!

If it appears that a fawn has been truly abandoned, contact your local wildlife rehabilitation refuge. If you're unsure how to find such a place in your area, just contact your local police. Do not try to take the fawn into your home and feed it milk, which will most likely cause diarrhea—and may even cause death.

## *Lyme Time*

The most common tickborne disease in the country, Lyme disease was first recognized in the United States in 1975 after a mysterious outbreak near Lyme, Connecticut, according to the Center for Disease Control and Prevention and the National Center for Infectious Diseases. The disease has since become a health concern in some areas of the country. The Humane Society of the United States reports more than ten thousand cases per year. There were a reported thirteen thousand new cases in 1994—an increase of 60 percent from the previous year. In 1996 the Center for Disease Control reported more than sixteen thousand cases. (To put these numbers in perspective, in 1982 the center reported slightly less than five hundred cases.) The huge increase is due, in part, to an increased awareness of the disease: More people knew about it, so more people got checked, and more were diagnosed. Primarily a problem in the eastern United States, the Upper Midwest, and California, Lyme disease has now been reported in forty-five states as well as the District of Columbia. New Jersey currently leads the nation in new cases of the disease.

Three conditions in nature need to exist in order for Lyme disease to pose a danger to an area: the Lyme disease bacteria, ticks that can transmit the bacteria, and mammals to provide food for the ticks.

White-tailed deer carry *Borrelia burgdorferi*—the bacterium that causes Lyme disease—in its adult stage. The tick that transfers the disease from mammal to mammal is usually referred to as a deer tick, but is actually even more prevalent in the deer mice found in the northeastern and north-central United States. (Another host of the Lyme disease-bearing tick is the white-footed mouse, which carries the disease in its larval and nymphal stages. Other mammals such as cats, dogs, cows, foxes, raccoons, skunks, and rabbits can also host ticks, as can birds.)

Two species of ixodid ticks are the primary carriers of Lyme disease: *Ixodes scapularis* (formerly known as *I. dammini*) in the East and Midwest and *I. pacificus* in California. Caribou and elk are confirmed carriers of another disease called brucellosis, but their role in transmitting Lyme disease remains unclear. Ixodid ticks are much smaller than common cattle and dog ticks. In their larval and nymphal stages they are no bigger than a pinhead, and adult ticks are only slightly larger.

The nymph must be attached to a human (or any other host) for at least twenty-four to forty-eight hours before infection can occur. The *Borrelia* that reside in the tick's midgut migrate to its salivary gland and enter the human's blood as the tick feeds. Usually there is an incubation period of three to thirty-two days before the clinical disease develops.

Most cases of Lyme disease are acquired in the early summer, when ticks are the most abundant and humans come into frequent contact with them. Ticks can only crawl; they cannot jump or fly.

They sit on grasses and shrubs, not trees, and are transferred to their hosts—either humans or animals—when these hosts brush against them.

If reading about how ticks suck blood (sorry, I gave the ending away) makes you queasy, skip the next section.

Ticks insert their mouthparts into their hosts, slowly feeding for several days as their bodies enlarge. The Centers for Disease Control and Prevention, National Center for Infectious Diseases, reports that in theory Lyme disease could also spread through blood transfusions or other contact with infected blood or urine, though no such transmissions have been documented. There is no documented evidence that a person can get Lyme disease from the air, food, or water used by an infected animal, or from sexual contact with an infected person. There is also no convincing evidence that Lyme disease can be transmitted by other animal-biting insects, such as flies, fleas, or mosquitoes. These insects transmit their own diseases.

Lyme disease is indicated by a characteristic ring-shaped rash—known as *enrythema migrans* and looking like a red bull's-eye—and flulike symptoms, which can include fatigue, chills, aches, fever, and headaches. These typically appear within ten days of infection.

Lyme disease is often difficult to diagnose because its most common symptoms mimic those of many other diseases. The fever, muscle aches, and fatigue can easily be mistaken for viral infections like influenza or infectious mononucleosis. Later-stage symptoms such as joint pain can easily be mistaken for arthritis, and still later neurological symptoms have been misdiagnosed as signs of multiple sclerosis.

Diagnosis for Lyme disease should take into account the possible exposure to ticks, the symptoms and signs, and the results of blood tests used to determine whether a patient's blood contains antibodies to Lyme disease bacteria. In early 1999 the Food and Drug Administration approved a test for Lyme disease that can be performed on the spot in a doctor's office. The test is called PreVue and is made by Chembio Diagnostic Systems of Medford, New York. PreVue was able to detect Lyme disease in 72 percent in a group of 120 infected blood samples and 95 percent in another group of 42 infected samples.

Lyme disease is rarely fatal, and the disease is easily cured if caught in the early stages by treatment with tetracycline, an antibiotic usually administered orally but, in severe cases, sometimes given intravenously. Left untreated, however, it can lead to serious complications. Untreated cases progress as blood flows throughout the body into the heart, bones, muscles, and internal organs. Lyme disease affects the skin, musculoskeletal, and nervous systems. Arthritis develops in the large joints, particularly in the knees. In the nervous system it causes memory loss, drowsiness, and behavioral abnormalities.

One of the best ways to avoid Lyme disease is to keep your body covered, especially when traveling through woodlands or places with heavy grass. In May, June, and July, if you're hiking or camping or even going for a short walk through woodlands, always wear long pants and a long-sleeved shirt, tuck pants into socks, and inspect your clothes and body afterward for ticks. Tick repellents, or insect repellents such as deet, worn on the body are also helpful, and you can treat clothes with permethrin, a pesticide approved by the FDA for use in contact with human skin.

The FDA has approved a vaccine against Lyme disease, but it's not entirely effective and is short lived. The best weapon against Lyme disease is prevention.

If you find a tick on your skin, remove it without crushing it and then thoroughly scrub the wound. Several outdoor companies and many drugstores sell tick-removal kits, which help you get the tick off before it infects your blood. If you believe you've been bitten by a tick—especially if you live in an area with a high rate of Lyme disease or any other tickborne disease, such as Rocky Mountain spotted fever—see your physician right away. Lyme disease isn't something to fool around with.

Flea and tick collars are highly recommended for your pets, especially if you live in or near a heavily wooded area.

You can also hire a licensed pest control expert to apply an acaricide, a chemical that is toxic to not only ticks but also your gardens, lawns, or fields—but be aware that some environmentalists have questioned the safety and effectiveness of such treatment. Pesticides can be toxic to humans and pets.

Some states are developing ingenious ways to kill ticks, one of which is a three- by five-foot deer-feeding station with vertical rollers on it—much like a car wash—that applies a coating of pesticide to any deer that walks the path to reach the corn stored in the feeder. In 1998 in the Washington, D.C., area, U.S. Department of Agriculture officials were recruiting volunteers to try out these bins in their yards.

Lyme disease is a major problem for a lot of communities, and understandably so. On the small, quaint Monhegan Island twelve miles off the coast of Maine, the seventy-seven residents were torn over a ballot question in the spring of 1998. Lyme disease had swept the island, forcing the inhabitants to make a painful choice—whether or not to kill the deer that had been roaming the land for more than forty years.

They'd arrived by ferry in the late 1950s, when state wildlife officials imported nine deer at the request of the residents to amuse tourists and provide game for hunters. Over the years, without any natural predators, the herd had grown to perhaps sixty deer— about forty-five too many for an island of only six hundred acres. The deer left the meadows and began to invade the tiny fishing village in search of food. They ate everything the islanders planted, bringing some wild vegetation to the edge of eradication. Fences did not seem to be able to stop the invasion.

Over just ten years eighteen year-round and summer visitors became infected with Lyme disease. A popular tourist attraction that has a population of more than four hundred visitors during the summer, Monhegan Island could not afford to maintain the status quo.

Various strategies were tried; none worked. Between 1994 and 1996 researchers fed deer ten tons of cornmeal mixed with ivermectin, an insecticide that makes female deer ticks ill during their breeding cycle, but the feeding and mild winters only caused the deer population to increase. In 1994 the islanders discussed injecting the deer with a contraceptive drug, but the idea never came to fruition. Because the deer population had grown exponentially, the villagers tried "culling," or selective hunting. A wildlife scientist and sharpshooter killed fifty-two deer over three days in 1997, and later in the year hunters were encouraged to take as many deer as possible—but this did not solve the problem, either.

Finally, the islanders faced the music. They hired a sharpshooter from New York who came equipped with a high-powered rifle, silencer, and night scope to annihilate the herd. End of problem.

Deer ticks that carry Lyme disease can also carry two other maladies: babesiosis and human granulytic ehrlichiosis. Though rare, these diseases will prove fatal in about 5 percent of humans who are infected.

Then there's tuberculosis, a serious respiratory disorder caused by several bacteria of the *Mycobacterium* genus. TB has three main types: human, avian, and bovine. Of these bovine TB is the most infectious, because it affects such a wide variety of wildlife— raccoons, coyotes, bobcats, opossums, red foxes, badgers, and deer. It can also affect humans.

Since tuberculosis is spread through the air, primarily by close contact with infected animals that cough or sneeze, the disease has a high rate of infection when unhealthy or stressed animals are crowded together.

As reported in *Deer and Deer Hunting* magazine in its November 1998 issue, 8 wild white-tailed and mule deer in Michigan were found to have had bovine TB in 1994, but a survey of 354 whitetails the following year revealed that at least 18 deer were infected with the disease. Surveys continued throughout the next two years, and by March 1998—after more than 8,600 deer were tested—149 individuals (or about 1.7 percent of the population) tested positive. Out of Michigan's eighty-three counties, however, only five were found to have infected deer. Since bovine TB threatens all domestic livestock, state officials must continue testing to ensure that there is no spread of the disease. Although only one cow was found to have contracted the disease, states such as Wisconsin and Virginia required testing of Michigan cattle before they were sold, an act that is estimated to have cost the state about $170 million over ten years.

## Road Warriors

Encounters with deer in gardens and backyards have led to some kicks and butts but seldom more serious harm to humans. Contacts between humans and deer on roadways are a different story. According to the Insurance Information Institute, a trade group in Washington, five hundred thousand automobile collisions a year are caused by deer. These accidents kill, on average, a hundred people per year, injure thousands more, and cost more than one billion dollars in property damage. The average deer-vehicle collision costs the motorist two thousand dollars.

In the 1930s approximately two thousand deer roamed the state of Tennessee. Today there are about eight hundred thousand deer. Frank Alexander of Dickson, Tennessee, told the *Tennessean* that he has killed three deer along the same stretch of highway. His most recent encounter ended up costing him $1,065 in repairs. "I was on my way to work . . . [and] the buck jumped into the road in front of the van, and I couldn't miss him," Alexander said.

After a game warden was accidentally fatally shot by a hunter who mistook him for a deer in 1993, Fairfax County, Virginia, banned hunting on most public land. Since then deer-human encounters have increased dramatically, occasionally with deadly results.

The following are a few examples of deer-versus-car incidents that occurred in and around the Washington, D.C., area, where the deer population is estimated at twenty-five thousand—about five times what the land can support:

- In July 1997 a paralegal hit a doe and was able to pull to the side of the road only after the deer had done more than four thousand dollars' damage to her van.
- In 1996 a deer wandered onto the Capital Beltway at Kensington, leaped toward an oncoming car, and crashed through the driver's sunroof. It landed upside down next to the driver and died. Luckily, the driver was only slightly injured.
- In October 1997 a librarian on her way to work was killed when a deer bounced off the fender of one car then crashed through the windshield of her Volvo.
- Nearby Montgomery County, Maryland, decided to combat the deer overpopulation problem and ensuing car collisions by conducting managed hunts, despite opposition from animal rights activists. Some property owners in the area have also begun to take the problem into their own hands by bringing in bow hunters to curb the herd on their own lands.
- Perhaps deer have political ambitions—dare we dream? Three does that took a wrong turn from Rock Creek Park found themselves caught in rush-hour traffic between the White House and the U.S. Treasury Building in downtown Washington. Traffic was snarled, though there were no injuries, as officials from several different federal agencies coordinated the roundup and relocated the animals to a more suitable environment.

It's hard to believe that you can die from deer, but if your car hits one, you can. What can you do to keep this from happening? In deer zones, drive no faster than the speed limit and always, always, always wear your seat belt.

Where there are trees, there are deer-car confrontations, Manhattan Island excepted. Skyline Drive cuts through Shenandoah

National Park in northwestern Virginia. A park of nearly two hundred thousand acres along the Blue Ridge in the Appalachian Mountains, it has amazing views of surrounding mountains. The area became a national park in 1935 and is lush with trees and teeming with wildlife, especially deer—one of which made a memorable impression on a driver, Tracy Quinn. She reports:

> I was with my best friend, Kristin, and we were taking one of our numerous day trips to nowhere. Part of the day's plan included my continuing effort to learn how to drive a five-speed car. Although I've been driving for more than a decade, I'd never had the chance or desire to drive a standard. But sometimes you've just got to challenge yourself.
>
> I took over the driver's seat once we'd made it past the entrance, and I began to drive. Feeling fairly comfortable with the fact that I didn't have to stop and start—cursed first gear!—I was able to cruise along at about fifty miles per hour—oh, how I love fifth—for about an hour or so.
>
> As I said, things were humming right along. I was checking the rearview mirror, looking at the picturesque scenery, and singing along to the Dave Matthews Band. Kristin had even started to doze off, I think. Things were going well until, lo and behold, a buck jumped into the road. Since I assumed he would continue crossing the road and I knew I couldn't stop the car in time, I decided to pull to the right-hand side of the road. The deer panicked, turned, and followed me. I was still going fairly fast and knew a head-on collision would have been a nasty sight (I won't even go into the time when my uncle hit a moose in Maine), so I pulled back into the middle of the road. Luckily, this time the deer didn't follow me.
>
> My advice for when a deer crosses your path: Swerve.

Virginia and Maryland are not the only states in America that are experiencing the tragic results when a deer meets the wrong end of a fender. More than half a million deer roam through Ohio—more than double the number of just ten years ago. In Ohio and Kentucky the number of deer-versus-vehicle accidents has quadrupled since 1980. In nearby Indiana animal-caused accidents have slightly increased each year since the early 1990s.

The North Carolina Department of Motor Vehicles estimates that from eight to nine thousand deer-vehicle collisions occur annually. In Iowa, the *Des Moines Register* estimates that there were eleven thousand such accidents in 1995, resulting in more than eleven million dollars in damages. In 1996 a deer on New Jersey's Garden State Parkway caused a ten-car pile-up.

What can we do to prevent these sometimes fatal and always costly encounters? Public announcements can be used to alert drivers to increased deer activity, especially during the autumn when deer begin mating. When bucks are in rut they often become aggressive and may even attack. If you see a deer by the side of the road you should never assume it will be timid or fearful; instead, be ready to react appropriately (by braking or turning to avoid a collision, as necessary) in case the deer should make any move toward your car.

The Humane Society also encourages highway and wildlife administrations to identify and post signs in spots where large numbers of accidents occur. Other approaches may include reducing speed limits in such areas, as well as removing vegetation from the sides of the roads so that driver and deer both have better visibility. Wood chips and stone along the edges of a road can be used to replace the grass that deer graze on. Fences that separate the woods from roads will also prevent deer from straying in front of a vehicle.

Bright lights save lives. Reflectors and well-lit edges on the sides of highways and roads are one way to prevent deer-vehicle collisions. The Strieter-Lite highway reflector system (see chapter 9, Resources, for contact information) consists of a specially manufactured plastic prism that is mounted along roadsides on steel fence posts and at prescribed distances, depending on road curvature and topography. The lights from oncoming vehicles at night reflect across the roadway into the area of woods or field where deer are, discouraging them from entering the road when vehicles are passing. Installation requires planning and approval through state and local Department of Transportation offices. Fairfax County in Virginia is just one area that thinks the reflectors are a good idea. The state announced plans in March 1999 to install reflectors along some county roads where cars often crash into deer.

Some further guidelines to help you avoid colliding with a deer:

- Keep a close watch for deer, especially at dawn and dusk or morning and evening rush hours on the road. About 20 percent of accidents occur in the early-morning hours, and 58 percent happen between 5 P.M. and midnight.
- Watch for DEER CROSSING signs.
- Drive with your headlights on bright and do not exceed the speed limit.
- Because deer prefer heavily wooded areas, be aware that they may be lurking behind the shrubbery lining the road.
- Exercise extra caution during the spring and fall, peak times for deer activity.
- Look out for a deer's eyes, which reflect light and glow.
- Because deer travel in family groups, if you see one cross your path watch out for another two or three behind it. If you see deer in the roadway, your best plan of action is to pull off the road, come to a complete stop, and be patient. Deer don't react like dogs do to a honking horn; they won't always run away.

If you do hit a deer or see an injured deer on the side of the road, report the accident to your local game authorities or call the police, who will contact them for you.

Living in northern Wisconsin, we encounter deer on the roadways quite regularly. On the opening day of deer season one year a deer ran onto the highway, crossing in front of my car and losing the battle. While driving a rental car two weeks later, waiting for repairs on car number one from deer number one, I was heading home and saw a young deer standing on the side of the road. I stopped, waiting for it to make the first move. It was like a staring contest, with nobody about to win soon. My patience ran out first, and I began to accelerate. I watched the deer run straight into my driver's-side door, turn, and run back into the woods.

Determined the following year not to repeat my experiences, I devised a theory that if I had honked my horn at the deer during the staring contest, it would have instantly run in the opposite direction, avoiding the arguments with the car rental agency about whose fault it really was that there was deer hair stuck in the door's trim (oops). I was given a chance to test the theory one morning on the way to church, with my husband driving. The roads were a little icy, and as we rounded a curve, standing in the middle of the road was a very large deer looking right at us and giving no indication of moving anytime soon. Concerned with the icy conditions, I yelled at my husband, "Honk! Honk the horn!" He honked and sure enough the deer began to run away from us, straight down the middle of the road, and then proceeded to fall *splat* right down on its belly with legs in all directions! It scrambled back onto all fours and then disappeared into the woods. Did I win or lose the battle? Well, the deer survived, and the vehicle was unscathed!

—Christine Reetz

# Outwitting Strategies from the Ground Up

*Successful gardening is not necessarily a question of wealth. It is a question of love, taste, and knowledge.*

—Vita Sackville-West

Gardening for pleasure is something thousands enjoy. But what happens when deer enter the picture? Deer, after all, are enemies—enemies of those of us who garden for the joy of it, for relaxation, for food, for something fulfilling to do while our significant other watches sports on television.

In the previous chapters I've tried to give you a sense of what we're up against. Now it's time to figure out if in fact deer are a problem for you. (If you bought this book because you *know* they're the problem, keep reading anyway. You'll learn something new.)

So you wake up one morning and find holes in your bushes in the shape of a small mouth. Uh oh. Some *thing* was in your garden. Was it your toddler? A vole? A deer? Somebody from the CIA opening a peephole into your yard? The fact is that deer won't spray paint your porch with THANKS FOR THE ROSES. NEXT SEASON, PLANT MORE! Still, there are ways to tell if they've tasted and perused your lovely handiwork.

Here's the very good news: You can garden and outwit deer at the same time, instead of erecting deer barriers *after* the critters have eaten your tulips. Sounds crazy, I know. But we are the species that created combination washer-dryers, computers that send faxes, cellular phones that are also pagers, and lots of other things that perform two (or more) functions at once.

Parents of small children are particularly good at performing several tasks at once, such as reading a newspaper, eating cereal, talking on the phone to the plumber, and spoon-feeding a six-month-old. If they can multitask on practically zero hours of sleep, so can you.

Think of the pride you will feel upon planting a beautiful garden, filled with colorful delights for your eyes and bitter, yucky booby prizes for deer. How's that possible? If you've ever drunk a glass of cod liver oil, you'll understand how deer will react when they arrive at your garden only to find that you have prepared their least-favorite foods. They'll be running back to their wild berries and forest ferns faster than you can say, "Bye-bye, bambinos."

## How to Tell If a Deer Has Been in Your Yard

### Basic Detective Work

Whether the damage to your garden or yard was done by a deer is not usually difficult to determine, especially if you've got an eyewitness report. It's not hard to catch them at it, either. Just wait patiently behind a bush or a tree, not making a sound, until you see them. Dusk and dawn are the best times to try. But if they continue to elude you, then go for the circumstantial evidence.

It's worth the effort and time to find out if your garden problem is caused by deer, digging squirrels, or some other animal. The techniques used to outwit one animal don't generally work for others, with one exception: big dogs. (Big dogs work for every kind of threat to your kingdom and are especially helpful if you also have an interest in outwitting your mail carrier.)

You'll know that you've had a deer nibbling on your saplings or stalks if they have a ragged, shorn appearance. Because deer don't have upper incisors, they cannot cut twigs as neatly as other animals. Bottom teeth touch a tough upper pad at the top of a deer's mouth. A deer grabs a mouthful of vegetation and shakes its head from side to side in order to pull it loose.

Rabbits, woodchucks, and horses tear vegetation off cleanly. Goats and llamas rip vegetation like deer. So if you live in Peru (South America, not Indiana) or your neighbor is a goat herder, your ripped tooth-mark evidence has not yet proved the case. The next thing to look for is whether there is a "browse line"—does the vegetation on trees and taller plants look chewed off to about three to five feet above the ground? Browse lines are formed when deer feed on vegetation as far as they can crane their necks. The line delineates the spot where deer can no longer reach the vegetation.

Deer can harm plants in more ways than browsing, though. From September to November, as the velvet dies from their antlers, bucks rub against small trees and shrubs to remove it. This aggressive rubbing helps prepare bucks to fight better and sharpens their antlers, but it's bad for the bushes and trees, and it also tears and tramples the ground.

## Deer Beds

You know a deer's made itself at home in your yard when you see a packed-down area of grass, roughly three feet in circumference, that functions as a deer bed. Deer usually locate their beds near their food source—in this case your garden. The deer logic appears to be: eat, sleep; eat, sleep; eat, sleep. . .

## The Tracks of My Deers

Look in mud, dirt, or snow for deer tracks, which are about three inches long. In most cases the hind foot of the deer will come down on top and slightly in front of the track made by the front foot. Remember, the deer's hooves are even toed. Cow and elk tracks are similar to those of the deer, but an elk's hooves are more pointed. Tracks of llamas and goats may also look like those of a deer, so know your area's wildlife.

Another hint: If you hear a lot of mooing, you don't have a deer problem.

## Scat, Another Four-Letter Word

Deer scat or droppings vary depending on the season. When the deer's diet consists mostly of browse (or the scarce scraps of vegetation available in off-peak seasons), usually in the fall and winter, the scat will be in the form of elongated, small pellets. In the spring and summer, when there is plentiful vegetation for deer to eat, their droppings clump together in a larger mass. You usually have to see the scat on the ground; once it's on the bottom of your shoe, you can't really tell what animal it came from.

## *Menace to Our Gardens and Livelihoods*

*USA Today* reported in 1998 that about twenty-five million deer are eating their way across our woodlands and grasslands, not to mention our parks, farms, and gardens. There are six times more deer today than there were in 1990—and that means many more problems.

Deer are becoming more tolerant of people, but the opposite isn't true. In fact, people are becoming less tolerant as the deer population increases.

Almost everyone who lives in an area with deer will tell you that they have had run-ins with them. Deer are hungry and have no manners. They don't think twice about trespassing into your yard for whatever they consider worth taking.

*Washington Post* columnist Mary McGrory writes: "I have always been for Bambi and against guns and blood sport. I took Princess Diana's side in her argument with Prince Charles about hunting. Lately, though, I've slipped some. I wonder if the Potomac Hunt Club would like to do a little tallyho-ing on Macomb Street [a street in urban D.C.]."

Damage to plants by deer has increased during the past decade, according to the University of Maryland. This increase is attributed to:

- rising deer populations
- human populations shifting to rural and suburban homesites
- loss of deer habitats due to development
- landowner decisions to prevent deer hunting

For deer, life revolves around the search for food, hence their appearance in your garden. They are more interested in eating than in anything else—more than television, spectator sports, or even sex. Deer eat more than five hundred kinds of plants. Each day an average-sized white-tailed deer eats about seven pounds of leaves, buds, blades, stems, berries, nuts, seeds, and other vegetation. In the winter their diet changes to mostly branch tips, small twigs, soft bark, and evergreen needles. An adult elk can eat up to thirty pounds of food daily, and a moose can eat up to fifty pounds. There are no Weight Watchers in the wild.

For whitetails, "candy crops" are acorns, apples, beechnuts, peas, and grapes, which means that planting any of these in your garden is like handing the deer an engraved invitation—unless, of course, you outwit them. They are also particularly fond of ground clover, alfalfa, berry briers, dried corn, and many other common garden plants. In October, when corn matures, deer will do anything they can to invade cornfields. If the naturally occurring crops they favor are plentiful, you may find your painstakingly planted garden less in need of vigorous defense.

Lichens, which grow on the bark of decaying trees, and wild mushrooms are favorites for whitetails in the fall and are especially prolific in damp years.

Deer eat mostly during low-light hours—at dawn and dusk—which helps protect them from possible danger. During a fullish moon you can see deer eating all night long. (The possibility of werewolves doesn't seem to scare them.) They also feed near cover where large vegetation shadows them. A deer is most vulnerable to its predators when it is out in the open foraging for food. Although deer tend to keep within their home range, they don't feed in one spot for too long and continue to rummage in order to limit their time in the open.

There are three kinds of gardens, all of them loved by deer. As suburbs spread into previously unoccupied land, deer are attracted to suburban gardens that encroach on their land. A rural garden is convenient to deer homes and most often unaccompanied by humans. Urban gardens are most often invaded during the spring, when young deer are chased away by their mothers in preparation for giving birth to new fawns. These young deer are inexperienced and often end up in community lawns and city parks, looking for handouts.

Generally, deer in the wild flee from human scent—unlike suburban deer, which are bold and don't hesitate to go into yards, gardens, and homes.

Deer often return to the same garden over and over again, as many people can attest. Once they have found their favorite foods (deer have distinct preferences, which are often individualistic) and learned that there's no danger, they will return to eat in their favorite spots. Deer know how to detect poisonous plants, either from natural instinct or from previous bad experiences.

And one warning: Starving deer become bolder. Droughts and other natural disasters can create a food shortage, causing deer to lose their inhibitions and go into territories they otherwise would not, eating plants they would otherwise have ignored.

Not only are deer a menace to hobby gardeners, but they are a danger to professional farmers as well.

The president of the New Jersey Farm Bureau, John Rigolizzo Jr., told the *New York Times* that farms in his state have lost upward of thirty million dollars in crops such as corn, hay, soybeans, string beans, strawberries, apples, pumpkins, and others to deer annually. The total harvest usually equals about $860 million. The New Jersey Division of Fish, Game and Wildlife estimates that there are about 150,000 deer living in the state, but Rigolizzo disagrees, saying there are more.

In eastern counties of North Carolina, deer have been called "crop destroyers." As a result, officials have permitted farmers armed with depredation permits to shoot and kill whitetails at any time in defense of their cash crops.

According to the U.S. Department of Agriculture, deer annually consume seven million dollars' worth of Iowa's corn crop.

Kentucky, which is estimated to have about half a million deer, has experienced an increase in crop-damage complaints. In 1997 these complaints rose for the second straight year, resulting in a rise in hunting permits. When the Kentucky Department of Fish and Wildlife Resources Commission set its 1997 hunting-season regulations, it liberalized the hunting zone in fifty counties with the goal of reducing deer populations to the benefit of farmers.

Dave Wingenfeld, a farmer in Valley View, Ohio, testified before the Cuyahoga Valley Communities Council that deer herds have eaten 20 to 55 percent of his corn, pumpkin, and pine tree crops annually. The Cuyahoga Valley National Recreation Area has begun implementing its goal of killing about five hundred deer between November and March to reduce the growing population, and the Cleveland Metroparks hoped to kill about one hundred deer on two of its reservations.

In a survey conducted in Virginia by the *Richmond Times Dispatch* and published in 1997, 70 percent of agricultural producers said they had experienced deer damage during 1995; totals ranged from $20 to $150,000 each. One in three homeowners said the same and estimated the cost of their damage between $20 and $1,000.

Vegetable farmers and cranberry growers in Massachusetts also report being plagued with hungry deer. The state estimates between eighty and ninety thousand deer—four times the size of the herd in 1970, according to the Division of Fisheries and Wildlife. In some South Shore communities the deer population is well above the preferred density level, which is twenty-six per square mile. Not only do deer feed on the cranberries, but their hooves tear up the bogs, leaving them damaged and adversely affecting the overall crop.

In West Virginia, the Christmas tree industry, which sells about thirty-three million trees each holiday season, is yet another casualty of hungry deer. Game officials estimate the state's deer population at about one million—with each animal looking for a free meal. After deer have eaten their way through the acorns, hickory nuts, and beechnuts, they are still desperate for more food. Christmas tree growers have been forced to replant thousands of seedlings that have fallen prey to deer, and they lose many adult trees due to browsing and rubbing.

## Planting to Outwit Deer

Okay, so you've determined that Bambi or Bruno has indeed been nibbling away at your prized petunias and precious peonies. Now what? I recommend a three-pronged attack:

- Avoid plants that deer like.
- Use plants that deer don't like.
- Design your garden to discourage deer.

It's that simple. (If only it were *actually* that simple, you wouldn't need this book.)

Deer have a wide-ranging diet, including thousands of plant

and vegetable species. However, deer generally don't eat plants with coarse, bristly, fuzzy, thorny, or spiny textures. *Washington Post* columnist Mary McGrory sadly came to the same conclusion at the end of a long, hard struggle with deer in her urban Washington, D.C., home:

> Bambi turns out to be mean: One night he leaves only a few trampled gaillardia behind. The next night he comes back and polishes them off. He passed on the impatiens— until, that is, I draped them in black netting, which he laughingly pushed aside. I cannot pave over the whole disaster with uncomplaining impatiens, always the gardener's last resort. No, now I have only one planting prospect—cactus.

But thorns aren't your only defense, and there are plantings that will do the job beside cactus. Fortunately for us, deer also dislike plants with intense aromas. The most important thing to keep in mind, however, is that it is still your garden—you need to be satisfied with the look and function of it.

Here's an overview of the list of tricks of the trade for planting that will help eliminate deer:

**Cut out the deer's browse line:** Trim off the lower branches of your trees in order to make them less attractive to deer. What deer will want to waste its precious time picking through your scarce yard if there are lush bushes next door?

**Block the deer:** If deer can't get into your garden, they can't eat. Fencing is the most effective antideer method, and the hardest to achieve.

**Protect the little ones:** Place small plants in the middle of larger ones, making them less vulnerable.

**Plant aromatic plants that confuse deer:** Since deer rely on their sense of smell to find food, you can hide some of their favorites by using strong-smelling plants to surround those most delectable to deer.

**Plant what deer don't like:** Substitute for your favorites similar-looking plants that have an intense aroma deer don't like. Your garden will look the same to you but won't be the happy eating place that it once was for deer!

**Block a deer's view:** Deer won't go over a barrier if they can't see the other side. Use shrubs, trees, fences, walls, anything you can think of to hide your garden.

**Make use of lawn ornaments:** Lawn ornaments distract a deer's attention from its goal: eating in your garden.

**Keep it neat:** Don't leave rotten fruit, acorns, or leaves on your lawn. You're just making a dinner plate for deer!

**Feed the deer:** Use plants such as alfalfa on the outskirts of your yard to keep deer full and happy, and they won't come into your yard.

Martha E. Wells tells me in an e-mail message:

> My own cornfield had to be planted about one and a half times as large as planned just so the deer wouldn't get the total harvest. Some older farmers say they always plant field peas in the last three or so rows for several reasons. These are "sacrificial rows." The peas don't grow as tall as the corn, the deer are nervous in open rows, and, theoretically, do less damage to the corn itself.

Jeffrey Chorba is a landscape designer from Beach Lake, Pennsylvania; he received his landscape design certification at the Morris Arboretum of the University of Pennsylvania. He also manages a computer service that provides landscape and horticultural information to landscape professionals and the general public.

Chorba advises people on designing for deer resistance—in other words, combining flora with fauna—in northeastern Pennsylvania. He has observed which plants tend to be deer resistant through all seasons. One thing he stresses is that resistant plants are those that are not, for the most part, bothered by deer under moderate conditions. In his opinion no plants are deerproof. He knows of plants proven as toxic to most animals, including deer, that were eaten by deer during extraordinary winters. With this in mind, he stresses, design and plant landscapes knowing that you will *always* experience some loss of plants.

Usually plants that are soft to the touch and high in water content are the first to succumb to deer damage. Chorba's experience confirms this; he has seen that deer usually like to eat the tender flower buds of plants first. Roses (*Rosa* spp.) and rhododendrons (*Rhododendron* spp.), if available, are the first to get eaten. As these food sources diminish, deer resort to less palatable plants, including forsythias (*Forsythia* x *intermedia*) and chokeberries (*Aronia arbutifolia*). He has found that, for the most part, plants with thorns or spiny projections turn deer away. Colorado spruce (*Picea pungens*), barberries (*Berberis* spp.), and hollies (*Ilex* spp.) are all perfect examples.

Flavor is another factor that can control deer browsing, according to Chorba. Plants with a strong taste will almost always survive a deer's visit. Marigolds are one example. Boxwoods (*Buxus sempervirens*) have also proved deer resistant due to their strong flavor.

A plant can have a certain degree of resistance because of its architecture. A good example is a crab apple (*Malus* spp.) tree (although deer will eat the apples if they can). If the deer can't reach the leaves due to the height of the canopy, the plant will thrive untouched.

Almost any type of ornamental tree can be used without concern for deer damage. The most important thing is that the tree should not be low branched, because low branches are easily reached and can detract from the overall appearance if they are damaged.

There are some deer-resistant shrubs that Chorba relies on to make a statement in the landscape. Through color, texture, and habit you can create many different effects even with a limited selection of plants. Red Japanese barberry (*Berberis thunbergii* Atropurpurea) is a great low-growing plant that can be used in a border or to add that splash of color to a rock garden. Inkberry *(Ilex glabra)* is one of the toughest, most versatile plants to use in the landscape. Chorba prefers to use the cultivar 'Shamrock'. This inkberry stays short cropped and keeps its foliage density even with age, unlike the straight species. It is a broad-leafed evergreen with rich, dark green foliage that stays free of insect and disease problems year-round. If you need a handsome evergreen specimen, a Swiss stone pine *(Pinus cembra)* is the perfect addition to your landscape. Swiss stone pines grow about eight inches a year and maintain a tight, pyramidal growth habit. Mountain pieris *(Pieris floribunda)*—sometimes called mountain andromeda—is an excellent massing plant that has creamy white flowers in the spring. This broad-leafed evergreen can be difficult to grow under some conditions; it works best in a slightly acid soil and a northwest exposure. It absolutely dislikes alkaline soil and full sun. Mountain pieris seems to work best in naturalistic plantings that have a good loamy, woodland soil.

Spireas can offer a pleasant touch of color in a hot, dry garden. Chorba uses Vanhoutte spirea (*Spiraea* x *vanhouttei*) for large masses that can screen unwanted views and brighten up a dull corner of the garden. Vanhoutte is one of the tallest spireas you can plant. It displays white flowers for about two weeks in the early summer. Anthony Waterer spirea (*Spiraea* x *bumalda* 'Anthony Waterer') is a fine plant with lavender-colored flowers that bloom for at least three weeks. This spirea tends to bloom from mid- to late summer nonstop till frost. There are many other species of spirea as well. For the most part, all spireas are deer-resistant and very reliable plants.

Ornamental grasses can offer a special textural element that perseveres through four seasons. Miscanthus and Calamagrostis are two common and very reliable grasses available at most nurseries. In most cases grasses prefer to grow in sunny locations where the soil is well drained. Most grasses actually prefer poor soil conditions.

Experiment with a variety of plants in your area to see if they are bothered. Deer definitely differ from area to area in what they eat. Also, regardless of the palette of plants you have to design with, it is always possible to end up with an interesting landscape with four seasons of interest.

## Deer-Resistant Plants

As I've said, no plant is entirely deer resistant, just like no room in a house is entirely childproof and no computer program is bug-free. Indeed, some gardeners will tell you that deer will eat any-thing, including "deer-resistant plants." When deer travel, they nib-ble along the way. Perhaps your best bet is to experiment by introducing one or two of each supposedly "deer-resistant" variety of recommended plants to see which actually work with the species of deer in your area. Whatever you find uneaten, plant more of. Ultimately, create your garden with it, and you may well end up with a deer-free zone.

### Deer Eating Facts
(Remember, these are facts and there's nothing you can do about that, no matter how much you wish otherwise.)

- No plant is totally deer resistant.
- Damage to plants sometimes depends on their location. If a

deer can't get to your roses, it loses out on one of its favorite
dishes.

- Deer damage your garden just as much by trampling
through it as they do by eating their way through it.
- Deer browse as they travel, so any plant that they come into
contact with on their daily route is likely to become a meal.
- Deer feed most in the spring, when natural food may not be
readily available, and in late summer, when they must fatten
up for the winter. They also raid gardens more in periods of
dry weather; they're looking for plants that have high mois-
ture levels.
- You won't be able to find anyone in agreement on what
plants are deer resistant. Depending on the species of the
deer and the locale of your garden, some deer-resistant
plants won't work.

A few intrepid gardeners have learned the secrets of having a
garden full of plants they enjoy that, at the same time, deter deer.
Because gardeners tend to be generous, helpful souls (at least
that's been my experience), they are usually willing to share the
secrets of their success, when asked. Through interviews with
some of these gardeners, through visits to their Web sites, and
through research of my own into various government and aca-
demic sources, I've been able to compile a list of all plants recommended to garden enthusiasts as having some deer-deterring
quality.

**An important caveat:** Nothing works in every garden, with every
species of deer. Before buying a new plant, be sure to consult
with a gardening expert (your local plant nursery should have
one) about its suitability to your soil conditions, climate, available planting space, and other considerations particular to your
own garden. You can save yourself some grief by talking with
other local gardeners and gardening stores before planting a

particular crop, too. Plant only in the correct season for that plant, to the depth indicated, and water and weed as that species requires. No plant will work to deter deer if it dies shortly after planting due to improper care.

## Vegetables, Fruits, and Some Herbs Deer Dislike

What follows is a list of vegetables, fruits, and herbs that deer won't eat either because of the smell or the texture.

| Botanical Name | Common Name |
| --- | --- |
| *Allium schoenoprasum* | Chives |
| *Allium* spp. | Onion, leek, garlic |
| *Allium tuberosum* | Garlic chives |
| *Arabis* spp. | Rock cress |
| *Arctostaphylos uva-ursi* | Bearberry |
| *Artemesia dracunculus* | Tarragon |
| *Asimina triloba* | Pawpaw |
| *Asparagus officinalis* | Asparagus |
| *Berberis* spp. | Barberry |
| *Cucurbita* spp. | Squash |
| *Cynara scolymus* | Globe artichoke |
| *Diospyros virginiana* | Persimmon |
| *Elaeagnus angustifolia* | Russian olive |
| *Ficus carica* | Edible fig |
| *Foeniculum vulgare* | Fennel |
| *Gaultheria shallon* | Salal |
| *Helianthus tuberosus* | Jerusalem artichoke |
| *Juglans regia* | Walnut |

| Botanical Name (Continued) | Common Name (Continued) |
| --- | --- |
| *Mahonia* spp. | Oregon grape varieties |
| *Myrtus communis* | Myrtle |
| *Ocimum basilicum* | Sweet basil |
| *Opuntia* spp. | Prickly pear |
| *Origanum pulchellum* | Marjoram |
| *Origanum vulgare* | Oregano |
| *Petroselinum crispum* | Parsley |
| *Phoenix dactylifera* | Date palm |
| *Punica granatum* | Pomegranate |
| *Rheum x cultorum* | Rhubarb |
| *Rosmarinus officinalis* | Rosemary |
| *Salvia* spp. | Sage |
| *Sambucus canadensis* | Blueberry elder |
| *Sambucus racemosa* | Red elderberry |
| *Solanum tuberosum* | Potato |
| *Sorbus* spp. | Rowan |
| *Tanacetum balsamita* | Coriander |
| *Thymus* spp. | Thyme |
| *Vicia faba* | Broad beans |

## Plants Rarely Damaged

This is the ultimate list, a combination of deer-resistant plants from a variety of sources in a number of places. I've included shrubs, trees, vine covers, climbers, perennials, and biennials. So as to preclude any garden store clerk from making fun of your pronunciation of the Latin names, I've listed the common names as well.

| Botanical Name | Common Name |
|---|---|
| *Abelia* spp. | Abelia |
| *Abies concolor* | White fir |
| *Acer rubrum* | Red maple |
| *Acer saccharinum* | Silver maple |
| *Acer saccharum* | Sugar maple |
| *Achillea* spp. | Yarrow |
| *Aconitum* spp. | Monkshood |
| *Agapanthus africanus* | Blue lily-of-the-Nile |
| *Agave americana* | Century plant |
| *Ageratum houstonianum* | Ageratum |
| *Ailanthus altissima* | Tree-of-heaven |
| *Ajuga reptans* | Carpet bugle |
| *Aloe* spp. | Aloe |
| *Amelanchier alnifolia* | Serviceberry |
| *Amelanchier laevis* | Allegheny serviceberry |
| *Amorpha fruticosa* | False indigo |
| *Amsonia tabernaemontana* | Amsonia |
| *Anemone* spp. | Anemone |
| *Angelica archangelica* | Angelica |
| *Anthemis tinctoria* | Golden marguerite |
| *Antirrhinum majus* | Snapdragon |
| *Aquilegia* spp. | Columbine |
| *Aralia elata* | Aralia |
| *Araucaria araucana* | Monkey-puzzle |
| *Arctostaphylos manzanita* | Manzanita |
| *Aristida stricta* | Wiregrass |

| Botanical Name (Continued) | Common Name (Continued) |
|---|---|
| *Artemisia* spp. | Wormwood |
| *Artemisia tridentata* | Basin sagebrush, big sage |
| *Arundo donax* | Giant reed |
| *Asclepias tuberosa* | Butterfly weed |
| *Astilbe taquetti* | Astilbe |
| *Aurinia saxatilis* | Basket-of-gold |
| *Beaucarnea recurvata* | Ponytail, bottle palm |
| *Begonia* x *semperflorens-cultorum* | Wax begonia |
| *Bergenia cordifolia* | Bergenia |
| *Betula* spp. | Heritage and river birch |
| *Betula papyrifera* | Paper birch |
| *Betula pendula* | European white birch |
| *Blechnum spicant* | Deer fern |
| *Boltonia asteroides* | Boltonia |
| *Borago officinalis* | Borage |
| *Brahea armata* | Mexican blue palm |
| *Buddleia davidii* | Butterfly bush |
| *Buxus sempervirens* | Common boxwood |
| Cactaceae | Cactus family |
| *Calendula officinalis* | Calendula |
| *Callistemon citrinus* | Bottlebrush |
| *Calluna vulgaris* | Heather |
| *Calocedrus decurrens* | Incense cedar |
| *Calycanthus occidentalis* | Western spicebush |
| *Campanula glomerata* | Bellflower |
| *Campanula medium* | Canterbury-bell |

| Botanical Name (Continued) | Common Name (Continued) |
| --- | --- |
| *Carex stipata* | Tussock sedge |
| *Carum carvi* | Caraway |
| *Caryopteris* x *clandonensis* | Caryopteria |
| *Catalpa speciosa* | Northern catalpa |
| *Catharanthus roseus* | Madagascar periwinkle |
| *Cedrus deodara* | Deodar cedar |
| *Celastrus scandens* | American bittersweet |
| *Centaurea americana* | Basket flower |
| *Centaurea cyanus* | Bachelor's-button |
| *Cerastium tomentosum* | Snow-in-summer |
| *Chaenomeles japonica* | Japanese flowering quince |
| *Chamaedaphne calyculata* | Leatherleaf |
| *Chamaerops humilis* | European fan palm |
| *Chelone glabra* | Turtlehead |
| *Choisya ternata* | Mexican orange |
| *Chrysanthemum weyrichii* | Chrysanthemum |
| *Cimifuga laciniata* | Bugbane |
| *Cistus* x *skanbergeii* | Rockrose |
| *Clarkia* spp. | Clarkia |
| *Clerodendrum thomsoniae* | Bleeding-heart (pink) |
| *Colchicum autumnale* | Colchicum |
| *Coprosma repens* | Mirror shrub |
| *Cordyline australis* | Dracaena palm |
| *Coreopsis grandiflora* | Coreopsis |
| *Cornus capitata* | Himalayan dogwood |
| *Cornus florida* | Flowering dogwood |

| Botanical Name (Continued) | Common Name (Continued) |
|---|---|
| *Cornus kousa* | Kousa dogwood |
| *Cornus racemosa* | Panicled dogwood |
| *Cornus sericea* | Red-osier dogwood |
| *Correa* spp. | Australian fuchsia |
| *Cortaderia selloana* | Pampas grass |
| *Corylopsis spicata* | Common winter hazel |
| *Corylus avellana* | Hazelnut |
| *Cosmos* spp. | Cosmos |
| *Cotinus coggygria* | Smoke tree |
| *Cotoneaster lacetus* | Cotoneaster |
| *Crocosmia* spp. | Crocosmia |
| *Crocus* spp. | Autumn-blooming crocus |
| *Cupressocyparis leylandii* | Japanese cedar |
| *Cupressus* spp. | Cypress |
| *Cydonia oblonga* | Quince |
| *Cytisus scoparius* | Scotch broom |
| *Dahlia* spp. | Dahlia |
| *Daphne* spp. | Daphne |
| *Datura inoxia* subsp. *quinque cuspida* | Thorn apple |
| *Delphinium* spp. | Larkspur |
| *Dennstaedtia punctilobula* | Hay-scented fern |
| *Dianthus barbatus* | Sweet William |
| *Dictamnus albus* | Gas plant |
| *Digitalis* spp. | Foxglove |
| *Diospyros virginiana* | Persimmon |
| *Echinops niveus* | Globe thistle |

| Botanical Name (Continued) | Common Name (Continued) |
|---|---|
| *Echium fastuosum* | Pride of Madeira |
| *Enkianthus campanulatus* | Redvein enkianthus |
| *Eranthis hyemalis* | Winter aconite |
| *Erica* spp. | Heath |
| *Eschsholzia californica* | California poppy |
| *Eucalyptus* spp. | Eucalyptus |
| *Eupatorium maculatum* | Joe-Pye weed |
| Euphorbiaceae | Spurge family |
| *Euryops pectinatus* | Euryops |
| *Filipendula rubra* | Queen-of-the-prairie |
| *Fraxinus velutina* | Arizona ash |
| *Fremontodendron californicum* | Flannel bush |
| *Fritillaria* spp. | Frittillaria |
| *Fuchsia magellanica* | Hardy fuchsia |
| *Gaillardia aristata* | Blanket flower |
| *Galanthus* spp. | Snowdrop |
| *Galium odoratum* | Sweet woodruff |
| *Gaultheria procumbens* | Creeping wintergreen |
| *Gelsemium sempervirens* | Carolina jessamine |
| *Gladiolus* spp. | Gladiolus |
| *Gleditsia triacanthos* | Honey locust |
| *Gypsophila paniculata* | Baby's-breath |
| *Hakea suaveolens* | Sweet hakea |
| *Hedera* spp. | Ivy |
| *Helianthus* spp. | Sunflower |
| *Heliotropium arborescens* | Heliotrope |

| Botanical Name (Continued) | Common Name (Continued) |
|---|---|
| *Helleborus* spp. | Hellebore, Christmas rose |
| *Hemerocallis fulva* | Tiger lily |
| *Hemerocallis* spp. | Daylily |
| *Hibiscus syriacus* | Rose-of-Sharon |
| *Hosta sieboldiana* | Plantain lily |
| *Humulus lupulus* | Hops |
| *Hydrangea* spp. | Smooth hydrangea |
| *Hydrangea paniculata* | Panicle hydrangea |
| *Hypericum calycinum* | St.-John's-wort |
| *Iberis sempervirens* | Candytuft |
| *Ilex* spp. | Dragon lady holly, San Jose holly |
| *Ilex cornuta* | Chinese holly |
| *Ilex glabra* | Inkberry |
| *Impatiens wallerana* | Impatiens |
| *Ipomoea purpurea* | Morning-glory |
| *Iris* spp. | Iris |
| *Jasminum* spp. | Jasmine |
| *Juniperus* spp. | Juniper |
| *Juniperus chinensis* | Chinese juniper |
| *Juniperus virginiana* | Eastern red cedar |
| *Kalmia latifolia* | Mountain laurel |
| *Kerria japonica* | Japanese rose |
| *Kniphofia uvaria* | Devil's poker, red-hot-poker |
| *Kolkwitzia amabilis* | Beautybush |
| *Laburnum* x *watereri* | Golden-chain tree |
| *Lamiastrum galeobdolon* | Archangel plant |

| Botanical Name (Continued) | Common Name (Continued) |
| --- | --- |
| *Larix decidua* | European larch |
| *Larix laricina* | Larch |
| *Lathyrus odoratus* | Sweet pea |
| *Laurus nobilis* | Bay or true laurel |
| *Lavandula angustifolia* | Lavender |
| *Leptospermum laevigatum* | Australian tea tree |
| *Leucanthemum maximum* | Shasta daisy |
| *Leucojum* spp. | Summer snowflake |
| *Leucothoe fontanesiana* | Drooping leucothoe |
| *Liatris* spp. | Gay-feather |
| *Ligustrum* spp. | Privet |
| *Limonium* spp. | Statice |
| *Linaria purpurea* | Toadflax |
| *Lindera benzoin* | Spicebush |
| *Linum perenne* | Blue flax |
| *Lithocarpus densiflorus* | Tan oak |
| *Lobelia erinus* | Edging lobelia |
| *Lonicera* spp. | Honeysuckle |
| *Lonicera caerulea* | Goldflame honeysuckle |
| *Lupinus* spp. | Lupine |
| *Lychnis coronaria* | Rose campion |
| *Lyonothamnus floribundus* | Catalina ironweed |
| *Lysimachia punctata* | Loosestrife |
| *Magnolia* x *soulangiana* | Saucer magnolia |
| *Matteuccia struthiopteris* | Ostrich fern |
| *Melaleuca quinquenervia* | Paperbark tree |

| Botanical Name (Continued) | Common Name (Continued) |
|---|---|
| *Melia azedarach* | Chinaberry tree |
| *Melianthus major* | Honeybush |
| *Mentha* spp. | Mint |
| *Mentha pulegium* | Pennyroyal |
| *Mentha spicata* | Spearmint |
| *Mirabilis jalapa* | Four-o'clock |
| *Mitchella repens* | Partridgeberry |
| *Moluccella laevis* | Bells-of-Ireland |
| *Monarda fistulosa* | Bee balm |
| *Morus* spp. | Mulberry |
| *Myosotis* spp. | Forget-me-not |
| *Myrica californica* | Wax myrtle |
| *Myrrhis odorata* | Sweet cicely |
| *Narcissus* spp. | Daffodil |
| *Nepeta cataria* | Catnip |
| *Nepeta mussinii* | Catmint |
| *Nerium oleander* | Oleander |
| *Nolina parryi* | Nolina |
| *Oenothera speciosa* | Evening primrose |
| *Oenothera tetragona* | Sundrops |
| *Onoclea sensibilis* | Sensitive fern |
| *Origanum laevigatum* | Origanum |
| *Osmunda cinnamomea* | Cinnamon fern |
| *Osmunda claytoniana* | Interrupted fern |
| *Osmunda regalis* | Royal fern |
| *Oxalis oregana* | Oxalis, redwood sorrel |

| Botanical Name (Continued) | Common Name (Continued) |
| --- | --- |
| *Paeonia* spp. | Herbaceous and tree peonies |
| *Papaver nudicaule* | Iceland poppy |
| *Papaver orientale* | Oriental poppy |
| *Parthenocissus quinquefolia* | Virginia creeper |
| *Pediocactus* spp. | Hedgehog thistle |
| *Pelargonium* x *hortorum* | Zonal geranium |
| *Petunia* spp. | Petunia |
| *Philadelphus coronarius* | Sweet mock orange |
| *Phormium tenax* | New Zealand flax |
| *Picea abies* | Norway spruce |
| *Picea glauca* | White spruce |
| *Picea pungens* | Colorado blue spruce |
| *Pieris japonica* | Japanese pieris |
| *Pinus mugo* | Mugo pine |
| *Pinus nigra* | Austrian pine |
| *Pinus resinosa* | Red pine |
| *Pinus rigida* | Pitch pine |
| *Pinus sylvestris* | Scotch pine |
| *Platycodon grandiflorus* | Balloon flower |
| *Plumbago auriculata* | Plumbago |
| Poaceae family (Gramineae family) | Ribbon grass (especially) |
| *Polemonium reptans* | Jacob's-ladder |
| Polypodaceae | Ferns |
| *Potentilla* spp. | Cinquefoil |
| *Potentilla fruticosa* | Bush cinquefoil |

‌‌‌‌‍‌‍‌‌‌‍

| Botanical Name (Continued) | Common Name (Continued) |
| --- | --- |
| *Primula* spp. | Primrose |
| *Prunus* spp. | Sweet cherry |
| *Prunus caroliniana* | Carolina cherry laurel |
| *Prunus serrulata* | Japanese flowering cherry |
| *Pseudotsuga menziesii* | Douglas fir |
| *Pteridium aquilinum* | Bracken fern |
| *Pulmonaria saccharata* | Lungwort |
| *Pyracantha angustifolia* | Fire thorn |
| *Pyrus calleryana* | Callery pear |
| *Quercus alba* | White oak |
| *Quercus muehlenbergii* | Chestnut oak |
| *Quercus rubra* | Northern red oak |
| *Quillaja saponaria* | Soap-bark tree |
| *Ranunculus flabellaris* | Buttercup |
| *Rhamnus cathartica* | Common buckthorn |
| *Rhododendron carolinianum* | Carolina rhododendron |
| *Rhododendron maximum* | Rosebay rhododendron |
| *Rhus ovata* | Sugarbush |
| *Rhus typhina* | Staghorn sumac |
| *Robinia pseudoacacia* | Black locust |
| *Romneya coulteri* | Matilija poppy |
| *Rosa multiflora* | Multiflora rose |
| *Rosa rugosa* | Rugosa rose |
| *Rudbeckia californica* | Coneflower (purple) |
| *Sabal blackburniana* | Hispaniolan palmetto |
| *Salix* spp. | Willow |

| Botanical Name (Continued) | Common Name (Continued) |
|---|---|
| *Salix matsudana* 'Tortuosa' | Corkscrew willow |
| *Santolina chamaecyparissus* | Lavender cotton |
| *Saponaria ocymoides* | Soapwort |
| *Sarcoccoca hookerana* | Dwarf sweet Christmas box |
| *Sasa palmata* | Bamboo |
| *Sassafras albidum* | Common sassafras |
| *Schinus molle* | California pepper tree |
| *Senecio cineraria* | Dusty-miller |
| *Sequoia sempervirens* | Redwood |
| *Solanum* spp. | Nightshade |
| *Solidago* spp. | Goldenrod |
| *Spartium junceum* | Spanish broom |
| *Spiraea* spp. | Spirea |
| *Spiraea* x *humalda* 'Anthony Waterer' | 'Anthony Waterer' spirea |
| *Spiraea prunifolia* | Bridal-wreath spirea |
| *Stachys byzantina* | Lamb's-ears |
| *Syringa* spp. | Persian lilac, Japanese tree lilac, late lilac |
| *Syringa vulgaris* | Common lilac |
| *Syzygium paniculatum* | Australian brush-cherry, eugenia |
| *Tagetes* spp. | Marigold |
| *Tanacetum parthenium* | Feverfew |
| *Tanacetum vulgare* | Tansy |
| *Tecomaria capensis* | Cape honeysuckle |
| *Teucrium fruticans* | Dwarf bush germander |
| *Thuja arborvitaes* | Cedar |

| Botanical Name (Continued) | Common Name (Continued) |
|---|---|
| *Tilia americana* | Basswood |
| *Tilia cordata* | Littleleaf linden |
| *Trachycarpus fortunei* | Windmill palm |
| *Trillium* spp. | Trillium, wake-robin |
| *Tsuga* spp. | Eastern hemlock, Carolina hemlock |
| *Tulipa* spp. | Tulip |
| *Vaccinium macrocarpon* | Cranberry |
| *Verbascum* spp. | Mullein |
| *Veronica* spp. | Veronica |
| Viburnum carlesii | Korean spice virburnum |
| Viburnum plicatum | Double file virburnum |
| Vitis labrusca | Labrador grape |
| Wisteria floribunda | Japanese wisteria (poisonous) |
| Weigela florida | Weigela |
| Yucca *spp.* | Yucca |
| Zantedeschia aethiopica | Calla lily |
| Zauschneria *spp.* | Zauschneria, California fuchsia |
| Zinnia *spp.* | Zinnias |

## Plants Frequently Damaged

In addition to knowing what plants will help to keep the deer away, it's also very useful to be able to avoid the reverse effect: plants that attract deer like flies. So unless you want your garden to send a mixed message (some plants scream, "Bambi, go home!" while others have that "come-hither" look), you should definitely eliminate these deer delights. Remember, if you have things that deer like and then you don't any more (because the deer have eaten them or because you've changed your mind), the deer will be used to your "restaurant" and come back—perhaps to try some other so-called deer-resistant plants.

| Botanical Name | Common Name |
| --- | --- |
| *Abies balsamea* | Balsam fir |
| *Abies fraseri* | Fraser fir |
| *Acer platanoides* | Norway maple |
| *Carya* spp. | Hickory |
| *Cercis canadensis* | Eastern redbud |
| *Chamaecyparis thyoides* | Atlantic white cedar |
| *Clematis armandii* | Clematis |
| *Cucurbita pepo* | Pumpkins |
| *Euonymus alata* | Winged euoymus |
| *Euonymus fortunei* | Winter-creeper |
| *Fagus grandifolia* | Beech |
| *Gaillardia pulchella* | Gaillardia |
| *Hedera helix* | English ivy |
| *Hosta sieboldiana* | Hosta |
| *Koelreuteria paniculata* | Golden-rain tree |
| *Malus* spp. | Apples |
| *Medicago sativa* | Alfalfa |
| *Rhododendron catawbiense* | Catawba rhodoendron |
| *Rubus* spp. | Blackberry brier |
| *Sorbus aucuparia* | European mountain ash |
| *Taxus baccata* | English yew |
| *Taxus cuspidata* | Japanese yew |

| Botanical Name (Continued) | Common Name (Continued) |
| --- | --- |
| *Thuja occidentalis* | American arborvitae |
| *Trifolium* spp. | Clovers |
| *Vaccinium macrocarpon* | Cranberry |
| *Zea mays* | Sweet corn |

From an Internet discussion group:

Question: "What are some deer-resistant plants for a rockery ground cover?"

Answer: "Many herbs and Mediterranean herbs such as lavender, santolina, thyme (wooly, red, etc.) are deer resistant. They have oils in their sap that deer don't like."

## Top Twenty Genera List

In compiling the long lists above, I noticed something important: The most successful deer-outwitting gardeners, *no matter where they lived,* named the same plants time after time. So I've made note of which plants were on everyone's best deerproofing list. I now present the Top Twenty Genera:

| Botanical Name | Common Name |
| --- | --- |
| *Achillea* | Yarrows |
| *Allium* | Onions |
| *Artemisia* | Sages |
| *Cedrus* | Cedars |
| *Hedera* | Ivies (except English ivy) |
| *Helleborus* | Hellebore |
| *Hydrangea* | Hydrangeas |
| *Juniperus* | Junipers |

| Botanical Name (Continued) | Common Name (Continued) |
|---|---|
| *Narcissus* | Daffodils |
| *Osmunda* | Ferns |
| *Papaver* | Poppies |
| *Pinus* | Pines |
| *Potentilla* | Cinquefoils |
| *Quercus* | Oaks |
| *Salix* | Willows |
| *Senecio* | Dusty-millers, bachelor's-buttons |
| *Syringa* | Lilacs |
| *Tsuga* | Hemlocks |
| *Yucca* | Yuccas |
| *Zinnia* | Zinnias |

## Camouflage Gardening

Another way to keep deer from eating your precious petunias is to practice camouflage gardening. It's really a three-step approach. The first part is to use plants that contain natural chemicals that deter deer from wanting to eat them; incorporate those plants found on the Rarely Damaged list into your garden landscape. The second step is to create a scent barrier. Since deer rely on their incredible sense of smell to determine what is good— and safe—to eat, a wide variety of strongly aromatic plants, herbs, and shrubs confuse the deer. If you've ever walked onto a public bus in the morning and encountered two dozen different perfumes—all at maximum strength—you'll have an idea of what this does to deer.

On its Web site, www.ghorganics.com, Golden Harvest Organics, Inc., advises gardeners to put a selection of deer-pleasing greens around the perimeter of the garden. This provides deer with something to eat, while protecting precious plantings. The company suggests gambel oak, four-wing saltbush, Rocky Mountain smooth sumac, Saskatoon serviceberry, and wood rose.

The third part of the approach to camouflage gardening is to plant deterrents that may stop deer in their tracks. Since deer won't venture past what they can't see, discourage them with a dense hedge or a trellis filled with flowers or plants (aromatic so that they're part of your scent barrier, and you're combining two efforts into one!). You can also create a deer barrier with a garden border using solid hedges or bushes, such as junipers and rugosa roses.

Carl Burgess, the owner of a 350-acre farm near Elk Garden, Tennessee, told the *Commercial Appeal* in Memphis, "I have a way of handling Mr. Deer. I plant decoy trees for him. He likes to get in the corner so he can hide and watch everything. So we plant some trees and let him chew them up. You have to know that fellow, know his habits."

Sara Shepherd tells me that she knows of several flowering plants, including hyacinths, that deer will not touch and in some cases won't go near. "Planting these around the perimeter of your yard or garden," she says, "is known to work and is usually nice to look at, too."

Judy Scott of Ithaca, New York, makes this suggestion: Use apples around the garden as a border. The deer will eat them and not harm your vulnerable plants.

## Plants with Strong Aromas

What follows is a list of plants, trees, and shrubs with strong aromas. Some of these plants serve double duty: not only does the strong aroma repel deer, but deer don't enjoy eating them anyway.

| Botanical Name | Common Name |
| --- | --- |
| *Achillea clypeolata* | Yarrow |
| *Anisacanthus quadrifidus* | Flame acanthus |
| *Artemisia cana* | Silver sagebush |

| Botanical Name (Continued) | Common Name (Continued) |
| --- | --- |
| *Buxus* spp. | Boxwood |
| *Catharanthus roseus* | Madagascar periwinkle |
| *Cedrus* spp. | Cedar |
| *Chrysanthemum leucanthemum* | Ox-eye daisy |
| *Cupressus* spp. | Cypress |
| *Dalea* spp. | Prairie clover |
| *Datura* spp. | Yellow datura |
| *Ericameria laricifolia* | Turpentine bush |
| *Flourensia cernua* | Tarbush |
| *Hymenoxys scaposa* | Four-nerve daisy |
| *Indigofera* spp. | Indigo |
| *Juniperus* spp. | Juniper |
| *Juniperus ashei* | Ashe juniper |
| *Juniperus virginiana* | Eastern red cedar |
| *Larrea tridentata* | Creosote bush |
| *Melissa officinalis* | Lemon balm |
| *Mimulus cardinalis* | Scarlet monkey flower |
| *Origanum vulgare* | Oregano |
| *Perilla frutescens* | Perilla |
| *Perovskia atriplicifolia* | Russian sage |
| *Phlox* spp. | Phlox |
| *Physostegia virginiana* | Fall obedient plant |
| *Pinus* spp. | Afghan pine |
| *Pluchea rosea* | Marsh fleabane |
| *Rudbeckia fulgida* | Black-eyed Susan |
| *Salvia apiana* | White sage |

| Botanical Name (Continued) | Common Name (Continued) |
| --- | --- |
| *Salvia azurea* | Blue sage |
| *Salvia farinacea* | Mealy sage |
| *Salvia greggii* | Cherry sage |
| *Salvia leucantha* | Mexican bush sage |
| *Salvia lyrata* | Lyreleaf sage |
| *Salvia regla* | Mountain sage |
| *Salvia roemeriana* | Cedar sage |
| *Santolina* spp. | Santolina |
| *Senecio cineraria* | Dusty-miller |
| *Sisyrinchium californica* | Golden-eyed grass |
| *Stylophorum diphyllum* | Celandine poppy |
| *Tanacetum vulgare* | Tansy |
| *Verbena bipinnatifida* | Prairie verbena |
| *Verbena peruviana* | Peruvian verbena |

## Get to Know Your Neighbor

Last, but certainly not least, is perhaps one of the most obvious places to go for gardening advice—the house next door. If you've just moved into the neighborhood or are only beginning to start planning your garden, consult people on your block or street about which plants seem to be deer-safe and which repellents have worked particularly well for them in the past. Just one or two conversations with the person next door can save you years of frustration and fights with our not-so-friendly Bambi. And after reading this book, you may have some useful tips for your neighbor, as well!

When outside gardening seems like too much of a hassle and

you're ready to give up, take the advice of A. J. Hicks of Socorro, New Mexico:

> I raise orchids, which are primarily tropical, and they grow inside. I have found that deer, like most people who enjoy *Wheel of Fortune,* are unable to work deadbolt locks and, in extreme instances, are incapable of using a common doorknob! Accordingly, this is how I outwit deer.

# Smarter than the Average Deer

*It is not enough to have a good mind; the main thing is to use it well.*
—Descartes

There are many ways to get rid of deer quickly and permanently, but for the time being, backyard nuclear weapons are still illegal.

I'll try to cover some of the tamer and more realistic strategies here. From the strangest repellents (human hair or urine) to the most obvious (fencing) to the ingenious (sonically activated high-pressure water cannons), people will try anything to keep deer out of their yards and gardens. (The next chapter will include information about deer-repelling products that are readily available over the Internet and through specialty catalogs.)

No single product or management technique, however, is likely to solve such a comprehensive problem as that of deer control. This is why I advocate a multifaceted approach, including vegetation management (chapter 3), fencing and repellents (this chapter), and population management (discussed in chapter 6).

Just as with overall techniques, it's worth keeping in mind that no single repellent works well all the time. Any one deterrent may keep the deer away for a short amount of time, but they may soon return once they realize that the smell of human hair, for example, isn't actually a predator lying in wait. Rotating deterrents on a regular basis will maximize their useful life.

Consider the season when you use repellents. Certain organic compounds can actually damage plants if sprayed on at forty degrees Fahrenheit or below. They tend to work best in the spring and summer when deer have more natural vegetation available to them. In the late fall and winter when there is little else to eat, deer have a powerful incentive—hunger—to overcome their fear or dislike of the repellent's effect to get the reward—the food in your garden.

## Fencing

Perhaps the most effective method of keeping deer out of your yard is by fencing them out. Deer-barrier fencing should be installed around the perimeter of the area you wish to protect. Because deer can jump incredibly high, generally a ten- to twelve-foot fence is needed, although some experts say you can get along well enough with a fence that's six to eight feet tall—if, that is, your local species of deer is on the smallish side. Low fences that block a deer's "landing zone" can also be useful, if well placed. (Go ahead and put your kid's stuffed wolf on the inside of your fence. Maybe the deer will leap over, see the wolf, then leap right back.)

It might be a comfort to know that deer are a problem in other countries, too. The Scottish Highlands, where they really know about deer, has its own problems. David Wright of Elgol, Isle of Skye, tells me, "It is not at all unusual to see twelve to eighteen red deer within fifty yards of the house when we get up in the morning, and there is plenty of evidence in the unfenced area that they come right up to the house during the night." Beautiful animals don't respect other people's plants, Wright testifies. "I am not aware of any reliably effective solution other than a six-foot wire fence such as surrounds our garden. Since our fence only surrounds the vegetable plot, we used rabbit netting rather than the usual sheep netting so that we could defend against two predators with the one barrier."

There are many kinds of fences, from homemade wooden models to high-tensile strand wiring to mesh-woven wire. Benner's Gardens has a polypropylene plastic-mesh fence that keeps deer out. It is a lightweight, high-strength fence that is virtually invisible and solves the deer problem without making your yard look like a maximum security prison.

David Benner, owner of Benner's Gardens, is a retired professor of ornamental horticulture living in New Hope, Pennsylvania. For years he tried everything he could think of to protect his garden from the local deer, to no avail. Wire, human hair, aluminum pie pans, and animal feces (more on this later) didn't work in the long run. Finally he developed the aforementioned fence, which he attached to trees in his yard. Enclosing more than two acres of his property with this fence, he found that the nearly invisible fencing not only kept deer out of his yard but didn't detract from the look of the land. (To find out more about this product, see chapter 5.)

Woven-wire fencing is a good option where the deer population is high and the likelihood for damage is great. This type of fence is usually eight feet high, constructed from two four-foot sections of six- by twelve-inch wire mesh joined with a hog ring. Two or more strands of barbed wire, spaced ten inches apart, are added to the top of the structure and extended to make the overall height ten or more feet. This fence must be routinely maintained and kept free of vegetation in order to be an effective deer barrier. Woven-wire fences are relatively expensive, especially if you take into consideration the cost of labor; they can add up to as much as four dollars per linear foot.

In places with few to no trees, pressure-treated wooden planks as well as fiberglass or metal posts can be installed to both support wire mesh and act as a fence themselves. The fencing is staked to the ground with metal pins every twelve feet to keep the deer from pushing their noses underneath it.

There are also standard chain-link fences and even electric ones.

High-tensile electric fencing has emerged as one of the best fencing methods. Relatively easy to erect, repair, and maintain, electric fences cost surprisingly little to operate because low-voltage chargers can be used to electrify long stretches—five thousand feet or more.

One proponent of electric fencing posted this note on the Internet:

> I live in an area where hunting is prohibited and many people feed the deer, so they are quite tame. I have discovered that deer will eat anything, including "deer-resistant plants." We have a six-foot fence around our acre and once the deer discovered there are tasty treats inside, the fence might as well not be there. They can jump much higher than my fence! We finally resorted to electric fencing around my garden areas, which takes away from the aesthetics but seems to be keeping the deer at bay. Put it about thigh high so the deer will lean against it to take a bite of what's inside. One touch is usually all it takes! Unknowing city folk think deer are so cute, but they quickly become pests when they destroy many dollars' worth of plants, not to mention the hard work put into a garden.

Fred and Polly Boggs agree that electric fencing is the way to go. They suggest:

> Put an electric fence around your garden. A wire three feet high will keep out most deer. Use four-foot for aggressive Bambi. Putting a second wire about four inches off the

ground will keep out most short-legged animals. . . . For extra protection hang a strip of aluminum foil on the top wire between posts.

One idea for attracting a deer to an electric fence is to put peanut butter smeared on aluminum-foil flags along it. Thus, the deer are initially attracted to the fence, but they soon learn that it is to be avoided and won't venture into your yard for fear of being shocked again.

Sound too harsh to you? A kinder, gentler barrier is the white-tape string fence. Hang the tape along the strings at varying heights, starting at about nine inches above the ground and ranging up to about forty-eight inches. The tape will look like a solid white wall to a deer, which sees a stationary object as a barrier. As an added bonus, the white tape will act as a pest repellent in the summer. While you're getting rid of deer, you'll discourage visits from mosquitoes, bees, and other pests with wings.

Slanted fences, those that lean toward the property to be protected, have had mixed success. A slanted fence is usually a five-foot-high, seven-strand, high-tensile, electrically charged fence that is tilted—or slanted—toward the garden at about a forty-five degree angle. Something about the three-dimensional, tilted tiers of these fences confuse a deer's depth of vision. This barrier uses flexible, spring-loaded, fourteen-gauge wire designed to survive the impact of a leaping deer. Deer seldom actually get into the property, but when one does, homeowners report, it's often with a broken leg incurred in the process of hopping the fence. Since having a wounded deer in your yard is worse than having a healthy, well-fed one, this type of fence is not something I would recommend.

Fencing poses its own problems, however. It may not be the most practical solution because of appearance, cost, terrain, or zoning restrictions. It's also hard to get into and out of your property if it's surrounded by a fence, especially if in your zeal to outwit deer you forget to install a gate.

We finally installed a critter fence at the edge of our property near the woods line. It's seven and a half feet high and has kept the deer out of the backyard. Now we put out block feed to watch the little angels. They come and bring their friends every evening at 6 P.M. sharp. Yesterday, I noticed that they seemed very interested in my daffodils [and I worry that] it's a matter of time before they learn to jump the fence.

—Sue Heller

If you do decide to use fencing as a deterrent, you should contact local wildlife experts to find out the right height and material for the type of deer in your area.

The cost of keeping deer out of your yard using fencing is between $185 and $5,000 per acre, depending on the type of fence.

To ensure that you don't have injured deer all over your yard after you install a fence for the first time, it is a good idea to put one-foot-long white streamers at the height of three to five feet (about the deer's browse line) at twelve-foot intervals at the top. These streamers don't need to be left up forever—only as long as it takes for deer to learn that there is a barrier. If a deer suddenly becomes frightened and does not see the fence and runs into it at full speed—up to forty-five miles an hour—it could damage both the fence and itself. Deer run-ins with fences are not common; they tend to occur only after a deer has managed to get into a yard through an unprotected access point and then, in panic, tries to escape through the fenced-in part.

The major deterrent to using fencing is that your yard will end up looking like an armed camp or battle zone, and that's an aesthetic consideration you need to consider before investing the money in this form of deer defense. But, perhaps the battle-zone look will become a much-imitated style.

In our area [Keno, Oregon], we have a crossbred mix of black-tailed and mule deer; they are larger than the blacktail and may be a bit more adventurous than some. They also seem to be less fearful of humans because they frequently browse in plain sight of people and have become accustomed to the smells. That apparently makes them less concerned about eating plants that are being protected by human scent of any kind (urine, hair, etc.).

I have used several of the commercial preparations with only limited success. Some have killed the new leaves on the plants and almost all have to be reapplied frequently. The deer sometimes stay away for a while and sometimes munch away right after I have sprayed. It seems to depend upon how hungry they are and/or whether there happens to be anything else around that they would rather eat at the moment. The anxiety I feel wondering whether the commercial preparations will decimate my plants hardly makes it worth it for me to use them, and my own recipes have about the same effect.

I have not tried the suggestion of fishing line. Maybe I just have a feeling that our deer would surely snicker at the suggestion that it would keep them out. Fencing is about the only thing that has worked for me. In this area an eight-foot fence on level ground or a ten-foot fence on a slope is recommended, and believe me, these critters would and do easily jump anything shorter. A friend, however, who lives in a different section of the same state uses a simple four-foot fence with a hot wire on top and has great success. But she has different deer. Ours just take the jolt or clear it if they want to get at the plants.

The only surefire method that I have heard of besides fencing seems to be dogs. But they have to be outdoor dogs, and it is just too cold here in the winter for me to keep one outside all of the time.

—Sharon Davis

## Scents, Homemade and Otherwise

When deer don't like the smell of your yard, or if they sense there is or has been a predator nearby (that includes us humans), they won't enter. Remember, deer don't like us any more than we like them. (Probably less; after all, we're worried that our lettuce will be eaten, while deer are worried that *they* will be eaten.) In fact, they will leave if they smell us, and not only when we haven't showered for a few days. Deer won't come in if they can't smell any food, so scents strong enough to mask the smell of tasty vegetation are helpful in keeping them away.

Scents are used in two ways. They are either applied directly to plants in order to make them smell terrible to deer, or they are hung or placed in the vicinity of your garden or yard, producing an odor offensive enough to keep deer away.

### Soap

Scented soaps are worth a try—and even if they're not 100 percent effective, you may at least enjoy their fragrance. Go for the heavily perfumed kind you can buy at a fancy bath shop; the scent is stronger and will last longer than ordinary supermarket brands. To save a few bucks (if you'll excuse the pun), try using Irish Spring, Dial, or Dove. Soap can be placed in nylon bags or cheesecloth sacks. You can also put a hole in the middle of each bar, tie it with string or coil, and hang it on trees or other shrubbery, leaving the wrapper on. Soap bars have an effective range of one and a half feet in radius. Ed Nortons tells me, "I'm from Minnesota and also have a cabin in Wisconsin. The deer love the evergreen. . . . In a cheesecloth or simple mesh bag, hang old soap pieces from the tree. It worked." Ammonium soaps also deter deer, but may also keep you from wanting to spend any time in your own garden.

Kathy, posting a message for an Internet discussion group, writes:

We live in a rural area where there are lots of deer, and I see them quite often. I kept them out of my vegetable garden last summer by hanging unwrapped bars of soap in mesh bags from trees on each side of the garden. I also took aluminum soft drink cans, cut slits in them, bent them back enough to catch the breeze, and hung them from a stake with four arms. I put one on each side of my garden and one in the center. Something sure worked because after I did this I didn't see any more deer tracks in the garden!

Most of the time my husband and I welcome deer on our acreage. We love watching them and have certain regulars that we look forward to seeing in the yard. We don't have plants in our yard just for the sake of looks and then get mad because deer happen to like to eat them. However, since we do plant for habitat and deer are not the only creatures we like to have around, we have to protect our seedling trees and bushes from the browsers.

Obviously, a wire cage works the best to keep out not only deer but also rabbits. When we haven't been able to use wire caging or fencing, we have tried something a friend in Wyoming told us about: Irish Spring soap. Take a bar of Irish Spring soap (or Zest or any strong-smelling deodorant bar) and use a potato peeler to shave off pieces of the soap onto the ground around trees or bushes or into an area you want the deer to avoid. I have even shaved some into the branches of evergreens to keep the deer from nibbling. It also worked when my husband got fed up with poachers and put the soap shavings in the deer path under the poachers' deer stand. It was obvious in the snow that the deer purposefully avoided this path and walked way out around that particular tree. Of course, the soap shavings have to be reapplied regularly, but they do seem to work and don't seem harmful to the environment.

—Mary Clark

## Other Stuff That Smells Bad to Deer

Some of the best deer repellents are not exactly things you find around the house. Bear droppings, for example. But if it works—and keeps working—it's worth hunting around for a source. The following products all work on the principle of convincing the deer that entering your territory would be too dangerous.

**Predator urine or feces:** The output of cougars, bears, bobcats, foxes, or wolves will activate the deer's instinctive reaction to flee at the smell of a nearby predator. The problem with "tankard" (as animal feces may be called) is finding them in a large-enough supply. I suggest you try contacting the circus whenever it's in town. Seriously. Most circus officials say that anyone with a bag, bucket, or pickup truck is more than welcome to haul away the manure from either the tiger or elephant cages; otherwise, the circus must pay a hauler to remove the stuff.

Tiger manure can be suspended from mesh bags and hung on wooden stakes at least thirty inches off the ground—or at the deer's nose level. Bags of predator manure may also be attached to rope fences around the perimeter of the garden. The bags will need to be refilled about once a month.

You can use a predator urine in your yard by simply soaking foam or another spongelike material with the product. (Predator urine commercial supply companies are listed in chapter 5.) The *Boston Globe* reports that word-of-mouth testimonials on the success of predator urine have increased sales at area companies from 17,000 bottles sold in 1992 to 250,000 bottles in 1997.

Urine is generally gathered from breeding farms where animals such as coyotes are being raised for their fur, to be put into zoos or game farms, or to be reintroduced into the wild. The predator urine is collected by rounding the animals into a large room that has drains on the floor. When the urine runs down the drain, it is filtered and stored in large drums for later sale.

Scientists in Colorado tested how different items repelled mule deer. Chicken eggs, Big Game Repellent, and coyote urine worked best because all are sulfur based. Predator urine and feces also have a high sulfur content because of the large amount of meat in the diets of these animals. (As the animal metabolizes meat, sulfur-based compounds are released.)

According to a study conducted in British Columbia, Canada, the most effective repellents for black-tailed deer are the urine of the cougar, coyote, and wolf (all common natural predators of the deer in Canada), followed closely by that of the bobcat and lynx.

In a 1991 study on white-tailed deer reported in the September 1998 issue of *Deer and Deer Hunting* magazine, researchers found that the urine of bobcats and coyote worked best to repel deer but discerned little response to rabbit or human urine.

On the other hand, Vaudeth Oberlander tells me that human urine worked just fine for him:

> I live in the Black Hills of South Dakota and am over-run by deer. Last year I started a new perennial garden and kept the deer at bay by pouring male human urine around the edges of the garden.

**Human hair** is another deer repellent, but, as deer become more accustomed to us, they are less prone to be scared away by our scent. As with soap, hang hair in sachets from trees around your yard. Hair (about two handfuls) should be placed in about one-eighth-inch mesh bags and hung approximately thirty to thirty-six inches from the ground. Hair should be placed at the perimeter of large trees at three-foot intervals. Clean hair is not as effective, so if you're planning on going to the local barbershop or salon for their clippings, which have just been shampooed, let the hair sit in the bag in your dusty garage or workshop for a few days before you hang it up, or try dumping in the dirt from your old vacuum cleaner bag. That should do it.

I used to like deer, before they started eating some pretty expensive hybrids I had planted. We moved here in August of last year, and I love flowers so one of the first things to do was to get plants in the ground for produce next year. I didn't know the area was feeding ground for three whitetails. They eat everything. To heck with the notion that there are plants deer don't eat 'cause there aren't any. Anyway, I kept trying things and the only thing I have found to slow them down is, and yes, I know this sounds gross, but I empty my vacuum cleaner bags out onto the flower bed! I even wrap some of it around certain plants to make a special effort to keep the four-footers off them. It has helped, though they may still nibble a little. At least they aren't eating the entire plant to the ground like they were.

—Laurie Klarman

Bernie Sayers tells me:

The easiest, most effective method to keep deer from a garden is to sprinkle human hair cuttings around the perimeter. Go to a barbershop and ask for their floor sweepings. Or if you cut hair at home, save it all. What I haven't determined is how long it works. Duration seems to be quite a few weeks but is related to rainfall. A more scientific test is needed to determine duration of protection. I learned this from local deer hunters. I thought it was nonsense until I tried it around some shrubs the deer were eating. I live smack in the middle of a two-thousand-acre game preserve with herds of deer everywhere.

Trisha Moore, another gardener, agrees:

The hairdresser I used to see to have my hair trimmed told me that she used to send all her hair trimmings to a guy in the Midwest, I think. He would put them around his gardens. Apparently, the human smell of the hair kept the deer away.

Deer also seemed deterred by the hair of other animals. Anita Jackson says, "I have a friend who uses dog hair strategically placed around the garden to deter all kinds of four-legged visitors."

**Blood meal and dried blood products**—which can be placed in cloth or nylon sacks and hung around your yard—prove a good deer deterrent because they, too, send the predator signal to deer.

Michael Geilich tells me:

> When growing things in the woods where I wasn't physically present a whole lot, I had problems with deer. I solved this by ringing each plant with a circle of dried blood; you buy it like fertilizer and it's actually deer blood. I also peed around each plant a lot.

**Other strong-smelling deterrents:** One good product that will help fertilize your plants as well as deter deer is called Milorganite, which is composted sewage sludge. You can buy it by the bag at your local garden store or from any of the sources listed in chapter 5. Hinder and Bobbex are two more commercial products that have scents particularly unattractive to deer.

Finally, here's a handy recipe that will send a deer away, bleating! Keep a quart of skim milk out of the refrigerator until the milk gets sour and lumpy. Add one-quarter cup of dish soap and shake the mixture. Spray it on the plants you want to protect. This mixture will need to be repeated at least every seven to fourteen days.

## Wire

Bob Tanem, who calls himself "America's Happy Gardener," recommends tightly strung piano wire installed in the shape of an X, using posts to support it. The bottom strand is strung twelve inches from the ground, the second at a three-foot height. For even larger deer, a third strand at a height of five feet is recommended.

Chicken wire and fishing line, as well as any other taut and nearly transparent wiring, will work equally well. One writer in the *Gazette* in Montreal lends support for this idea:

> In several experiments it's been shown that if you stretch clear monofilament fishing line across the path of a deer, they can't see it and when they touch it they freak out and run away. Wonderful! So I can see that our garden will have a series of fishing line perimeters around the beds and throughout the nearby woods.

In Maine they even use wire lobster traps to repel deer, but somehow I don't think that deer traps would work on lobsters.

## Scare Devices

Any sight out of the ordinary for deer will deter them, but don't rely on these scare tactics to work for long. As deer become accustomed to their surroundings and realize there's no danger, they will encroach on your property once again.

Scarecrows and other effigies may repel deer for a short amount of time, but they are usually ineffective in the long run. Other visual deterrents that may have long-lasting repellent effect include:

- fluttering metallic ribbons
- flags
- balloons
- cassette or videocassette recorder tape (removed from the cartridge)

These items need to be hung so that they will flutter or bob in the wind close by the plants you wish to protect.

Cass Peterson, as quoted by the *St. Louis Post-Dispatch*, reports encounters with all sorts of woodland scourges such as ground-hogs, crows, rabbits, voles, bugs, and slugs, but deer are the biggest enemy. Recommendation?

Scare tactics. Deer are jittery animals. The idea is to keep them spooked with strange noises or flapping objects. I've tried plastic trash bags tied to tomato stakes, soda cans strung up so they rattle in the breeze, and transistor radios turned to a heavy-metal station. All these methods work for three days or less.

## Lights

Some people have been known to hang strobe lights on a fence to repel deer, although the number that it would take in order to scare away a single animal probably isn't worth the cost to your electricity bill.

Lights that switch on via motion detectors may help protect gardens and yards better than bright lights because they are activated only when a deer enters your premises; the sudden change in lighting will usually scare it enough so it runs away. Like scarecrows, however, lights become ineffective once deer realize that no danger follows these surprises.

Perhaps the best use of a light is as a means to alert a homeowner that there is a deer—or something else even scarier—outside.

## Netting

Dacron netting, which can usually be found in seven- by one-hundred-foot units, can be a good deer deterrent. Drape it over trees and crops to keep animals away. Black plastic netting can be expensive, however. Instead, try a good cloth netting, which is usually sold in six-foot widths, from the fabric departments of outlets such as Wal-Mart or other discount stores for about ninety-nine cents a yard.

A see-through mesh netting called Invisible Mesh Barrier is also available at your local garden or home improvement center or through gardening catalogs (listed in chapter 5). But make sure to check with your local government, which may consider these mesh materials to be fences and thus regulate their height through their own zoning laws.

## Noisemakers

If you can stand it, anything that makes noise is likely to scare off deer. Strings of aluminum cans or pieces of old scrap metal suspended between trees and blowing in the breeze will frighten a deer.

Deer may also run from music, which you may find pleasant to hear on outdoor speakers or shower radio. (Just make sure your neighbors don't mind, if they live within earshot.) The same stuff that works for squirrels—Led Zeppelin and Frank Sinatra—seems to work best for deer.

One creative way of scaring deer involves a contraption that involves a big metal pipe on a hinge that balances like a teeter-totter. You rig up a hose to drip water into the top of the pipe; when the pipe gets full, it tips down and clangs on a rock, scaring the deer away.

Landscape designer Brian Crick tells more about this type of device in an Internet posting:

> It is a bamboo seesaw [with] one end intact without the membranes damaged, the other end cut at an angle with some membranes removed. This is set up like a seesaw with the natural balance being to the intact end, which sits on a rock. The open, angle-cut end is fed with a slow trickle of water. As it fills, the center of balance changes till it finally tips and the water flows out. Then it returns to the resting position with a *bang* as the bamboo hits the resting rock. . . . Apart from their historical use to scare away deer, they are a great attraction in any garden.

Wind chimes may also scare deer in the short run. Two more frightening options are gunfire and pyrotechnics, but unless you live on a police shooting range or have a license to test fireworks, these won't be practical solutions. (If you intend to shoot blanks or use a starter pistol, alert your neighbors first, or you may have some explaining to do to the police.) One less far-fetched option is the use of a propane cannon. These cannons must be moved every few days, and the sequence of explosions should be staggered so that deer will not get used to them. They are only effective for a couple of weeks, however, because deer eventually do become accustomed to them—by which time you'll already be sick of them. If, however, you think it's worth a try, then be sure to check with local officials about restrictions or regulations on the use of explosive devices.

Here's how Lynne Mellinger of Ithaca, New York, solved her deer problem:

> We just built a beautiful home in a former meadow near Cornell. We are surrounded by woods and ponds so, of course, our area abounds with animals of all kinds. Coming from Pennsylvania where the only "wild animal" we ever got to see was a runaway Amish horse, we are just in awe of our surroundings and enjoy it immensely.
>
> Among other animals, we found we had six to ten deer in a herd that run behind our house. Most people here dislike the deer because, of course, they eat everything. Because we built our home late in the year, last year I spent the winter deciding what to plant. I checked online information about deer and what they do and don't eat, found lists that suggested what I should plant and shouldn't plant, and went from there. We have a ten-foot by eighty-five-foot bank in back of our house where I decided to put a flower garden, and we were building a corner pond next to it because we have springs running all through it. With that much area to cover, I sure didn't want the deer to eat my posies.

I started out with shrubs in early April followed by some of the hardier perennials. Now, the lists and the nurseries say that deer don't eat astilbe, mountain laurel, or daylilies, but I'm here to say they do! I walked out one morning to survey the previous day's work to find some little nipper had forged a path of destruction through upper half of one side of the garden. Everything had at least a chomp or two taken out, and I was furious.

We had big spotlights trained on pine trees in back, which we turn on at night to watch the animals. Last winter I had a big pinecone stuffed with a peanut butter concoction hanging from the pine tree for the squirrels. The deer were terrified of this at night when they came around. They would not go near the corn on the ground if this happened to sway slightly in the wind.

So I bought a ten-inch wind chime, which has a nice sound and a small pole, and stuck this right in the middle of the astilbes at the top of the garden. . . . Well, the deer stayed away. I bought another larger chime for the other side of the garden, and the deer have never bothered anything over there since.

## Ultrasound Devices

Because experts say that deer find high-pitched sounds intolerable, most pest-control product catalogs feature ultrasound devices, which make noises that humans can't hear, with the assurance that they will scare away deer. Most are motion activated—they turn on only when someone or something crosses their path. However, there's a lot of disagreement about whether these devices are at all effective. If there ever were a case of buyer beware, this would be it.

Very important: Do not rely on ultrasonic whistles mounted to your car to prevent deer crashes. They don't.

Even if the ultrasonic noise is shown to repel deer, it is almost certainly the case that the effect will be temporary because the animals eventually discover that no harm follows upon hearing the unpleasant sound. Yet another drawback is the effect of the ultrasonic noise on neighborhood pets and livestock.

## Sprinklers

Sprinklers set at random intervals are a cheap and reliable method of scaring off deer. Better still are sprinklers that are motion sensitive. Heat-and-motion-activated sprinklers, such as the Sensor Controlled Animal Repeller, can be mounted in the garden and connected by a hose to your outdoor faucet. The sensors, which are powered by 9V batteries, send out a blast powerful enough to send a deer scampering, and as an added benefit help keep your lawn lush and green.

You don't have to spend a lot of money for a ready-made system. As gardener Janine Brennan notes, you can build one yourself from parts:

> After battling four-legged pests in my gardens, I have found a wonderful solution. I put together a system of motion sensors and sprinklers using off-the-shelf components from my local builder's supply store. This system works 100 percent in keeping deer, groundhogs, rabbits, etc., out of my gardens.

## Dogs

Having a dog is one of the surest ways to deter deer. Let your dogs run wild in your yard and you'll have happy pets and be virtually deer-free. Fence your property to make sure your dog stays there to guard it. He's a natural predator that deer will fear. Even if deep down he's a pussycat, there's always the smell of canine urine to scare the herds away.

Dogs win the endorsement of many online garden enthusiasts. A sampling follows:

> I haven't had a large garden in several years, but we do have large guardian dogs, Asian shepherds, who are not destructive to the yard area (and not "yappy") but are good deterrents to deer and coyotes.

> —Martha E. Wells

> We have many deer around us, but they leave the plants in my yard alone. We have 1.5 acres cleared around the house, except for some big oaks. I also have azaleas planted and at one time I had many hostas. The moles/voles got the hosta. I have neighbors who have lost azaleas and other small trees and bushes, even roses, to the deer, but we have not. The only thing I can figure is that we have a sweet old golden retriever that hangs around the house—outside most of the time. We have looked out the window in the winter and spring and seen deer on the edge of the lawn, but they have never eaten our bushes!

> —S. Perry

> When growing vegetables in the yard without a fence or any physical barriers to deer, I've found having a dog around keeps the deer away after one or two chasings. The real trick was teaching the dog to stay out of the garden.

> —Michael Geilich

> One solution to deer control I don't hear much mentioned is the use of dogs. I've a friend who lives on ten acres of mostly wooded land and has wonderful ornamental gardens surrounding his home. Never are they bothered by deer despite the fact that deer are rampant—way too many for the

land to support. He has two dogs of shepherd/husky mix, which are pets. They are quite lovable dogs but by allowing them the run of the property the deer are kept away.

—Steve Graf

And finally, a thought from Dennis Mathiasen, who's used lots of different techniques but tells why he thinks dogs and sprinklers work best:

> Deer won't, in general, hang around places inhabited by humans unless hunger has driven them to look for food there. With that high level of hunger they're not going to be very selective in what they eat, so the best ways to deal with them are to keep them out in the first place with fencing or use devices that work by scaring them away. Some examples are free-running dogs or water sprinklers connected to sensors. Anything that they can get used to won't continue to work for long.

There are some downsides to using dogs as a deer deterrent. The first is fleas. If you've never had them in your house, you've never understood the need for knee socks. Another is dog poop. Let a dog roam wild and it will poop where it wants to. Dogs also bark. While you may grow accustomed to your own dog's barking—and even find the sound "sweet"—your neighbors won't necessarily be so enthusiastic. (Also see the section in *Outwitting Critters* on dogs.)

two months, rain or shine! It keeps black-tailed deer, white-tailed deer, Roosevelt elk, and rabbits away from ornamentals, conifer seedlings, dormant fruit trees, and shrubs, yet is not offensive to humans.

Deer-Block Protective Mesh Fence
Phone: 1-800-327-9462
Easy Gardener Catalog
P.O. Box 21025
Waco, TX 76702-1025

Deerbusters
(a division of Trident Enterprises)
Phone: 1-800-248-DEER
9735A Bethel Road
Frederick, MD 21702-2017
www.deerbusters.com
E-mail: deer@deerbusters.com

Deerbusters products are designed to address deer damage and other wildlife problems in gardens, flowers, trees, shrubs,

*Reprinted with permission from Deerbusters™*

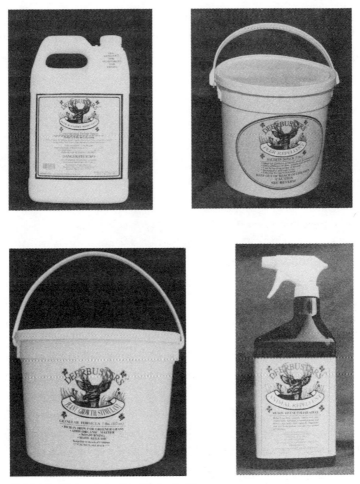

*Reprinted with permission from Deerbusters™*

and landscapes. The company provides a full line of deer fencing, including electric baited deer fences, nonelectric invisible mesh barriers, deer netting, deer repellents, scaring devices, products to control Lyme disease, and other pest-control products.

The company's Web site provides a deer-resistant plant list, among other interesting and informative factoids. Deerbusters and Dr. Deer will create a personalized Deer Management Program specifically designed for your needs that will not harm deer.

Deer No No
Phone and Fax: 860-672-6264
P.O. Box 112
West Cornwall, CT 06796
E-mail: deernono@mail1.nai.net
Deer No No has a specially formulated citrus scent that deer hate but that is pleasing to humans. The solid cake form lasts ten to twelve months. It comes in a net bag that can be hung on a branch or attached to a stake and stuck in the ground.

Deer-Off
Phone: 1-800-DEER-OFF
www.deer-off.com
An all-natural deer repellent made from biodegradable food products with no harmful chemicals, Deer-Off is safe for humans as well as the environment. It is registered with the Environmental Protection Agency. A patented dual-deterrent system leaves both an odor and a taste that deer find offensive. One application lasts three months and does not wash away in inclement weather. For best results, use Deer-Off when temperatures are above forty degrees Fahrenheit, there is little or no wind, and the leaves are dry.

Deer-Outahere
Phone: 605-341-6847
Black Hills Energy, Inc.
2120 West Main Street, Suite 7
Rapid City, SD 57702
A chemical-free, homemade deer repellent that is guaranteed

to work or your money back. So how can you go wrong? The company sends you a recipe that can be made with ingredients commonly found at home.

Deer Pro-Tec & Deer Ropel
Available through: Plow & Hearth Catalog
Phone: 1-800-866-6072
P.O. Box 500
Madison, VA 22727-1500
Pro-Tec Capsules were developed by a forestry researcher to take advantage of a deer's keen sense of smell, which is a thousand times more sensitive than a human's. The easily concealed yet effective capsules use garlic oil to overpower and disrupt other scents that attract deer. If a deer does nibble, the pepper oil in the capsule will change its mind.

Perhaps the worst-tasting substance known to deer, Ropel is not toxic but does discourage them from nibbling. Spray Ropel on your plants, but don't spray it on something you want to eat—you won't like it any more than the deer do!

FoodComm International
Phone: 1-800-445-4622
4260 El Camino Real
Palo Alto, CA 94306-4404
www.foodcomm.com
An online store for venison, duck, and other specialty meats. Also has venison recipes.

Gallagher Power Fence
Phone: 1-800-531-5908
P.O. Box 708900
San Antonio, TX 78270
When touched, these fences give a shock. Although high in

An environmentally safe way to repel deer (as well as squirrels, rabbits, and chipmunks), N.I.M.B.Y. is an emulsion of natural oils that comes in a thirty-two-ounce trigger spray bottle that sells for $14.95. One or two applications proved to last an entire season. N.I.M.B.Y. can be sprayed directly on plants; approval is pending from the Environmental Protection Agency for use on fruits and vegetables.

Not Here Deer
Phone: 1-888-NOT-DEER (1-888-668-3337)
P.O. Box 143
Armonk, NY 10504
www.notheredeer.com
E-mail: nhdinfo@cognizant.com

Not Here Deer, the only patented electronic device specifically designed to unobtrusively repel and disperse deer, reduces the damage caused by deer browsing.

Not Tonight Deer!
Phone: 415-255-9498
77 Waller Street
San Francisco, CA 94102
www.nottonight.com/core.html
E-mail: info@nottonight.com

Not Tonight Deer! is an all-natural deer repellent that can be used to protect fruit and vegetable gardens. Besides having a great name, this product keeps deer away using simple food products and no harsh chemicals. One six-ounce bag of powder covers over five thousand square feet and costs $9.99.

Things customers say: "One of the most effective and best-selling repellents in our store. It gets our highest recommendation."

—Jim Chiapelone of Burlingame Garden Center,
Burlingame, California

*Reprinted with permission from Not Tonight Deer!™*

"Keeps my vegetable garden and roses safe from the local deer herds and for that I thank you."

—Bob Lucas, Polsbo, Washington

"One of the few deer repellents that we found to really work."

—Julie Petrie of Breed & Company Nurseries, Austin, Texas

Peaceful Valley Farm Supply
Telephone: 530-272-4769
Fax: 530-272-4794
P.O. Box 2209
Grass Valley, CA 95945
www.groworganic.com

The Pepper Shop
www.ThePepperShop.com
E-mail: PepperGrow@aol.com

An excellent online source for hot pepper sauces, live pepper plants, tips, photos, facts, and even recipes! The Pepper Shop also provides links to other useful Web sites; click on the "pest control" link and chat with other people plagued by pests. Many folks have found success by using a combination of traditional deer repellent sprays, soap, and hot peppers.

100% Predator Urine
Phone: 1-800-218-1749 (8 A.M.-7 P.M. Eastern Standard Time)
Fax: 207-862-2267 (twenty-four hours a day, seven days a week)
J&C Marketing, Inc.
P.O. Box 125
Hampden, ME 04444
predatorpee.com/home.html

These Pee Products, as the company calls them, work by triggering the natural protective instincts of animals. The scent of predator urine creates the illusion that predators are nearby. Deer fear coyotes and wolves in particular; hence, these products work best to keep deer out of your garden.

The Predator Urine concept is based upon the principle of duplicating the behavior of predators in the wild, which mark the perimeter of their territory with urine to warn away competitors. Just place the company's Scent Darts into the ground every ten to twelve feet around the designated area, and squirt the appropriate product. (Predator Urine also offers a wide selection of urine to keep moles, raccoons, gophers, and other garden wreckers out of your yard.) Reapply weekly or after it rains.

Use the company's simple online order form or just call or fax your order to the number above.

What customers say: "Effective"; "most excellent"; "Coyote Urine is the most effective method I have ever used"; "thank you so much"; "worked wonders."

Romancing the Woods, Inc.

Phone: 914-246-6976

33 Raycliffe Drive

Woodstock, NY 12498

www.rtw-inc.com

E-mail: davis@rtw-inc.com

Romancing the Woods, Inc., sells rustic garden structures designed to make your backyard both deerproof and aesthetically pleasing. Eastern red cedar has an aromatic quality that repels deer, while the woodwork is influenced by eighteenth- and nine-teenth-century English and European estate garden designs. The company ensures that all work complies with local and state building codes. High-profile customers include Disney's Animal Kingdom (twelve benches) and Johns Hopkins University in Maryland (which purchased a twenty-five-foot bridge).

*Reprinted with permission from Martin Davis, Romancing the Woods*

*Reprinted with permission from Martin Davis, Romancing the Woods*

What customers say: "From the start, you treated me like I was your most important client"; "easy to work with."

Scarecrow
Phone: 1-800-767-8658
Fax: 1-800-876-1666
Contech Electronics, Inc.
P.O. Box 115
Saanichton, BC, Canada
V8M 2C3
www.scatmat.com

The patented Scarecrow claims to be the smartest scarecrow ever invented, boasting a customer satisfaction rating of 89 percent. When its high-quality motion sensor "sees" an intruder, it instantly sprays a garden hose full of water at high pressure at the trespassers. The effect is both startling and immediate! Animals quickly get out of the area and avoid it in the future. The secret

*Reprinted with permission from Contech Electronics, Inc.*

"I think I have finally found a way to keep deer from dining at Diane's. I bought a generic brand of Milorganite, which thus far seems to be working! It comes in little pellets, which I sprinkled throughout my garden and around the base of other plants [deer] like and so far they're still untouched. As well as keeping deer away, it is also fertilizing the plants and promoting root growth at the same time. The deer have gone so far as to step up on my front porch steps to nibble on the large bushes that flank the sides of the porch. . . . So far the pellets seem to be working better than our dog at keeping the deer away. We will be keeping the dog, just the same, and hoping we've rid ourselves of the deer dining at Diane's."

—Diane Manning

# Looking at the Big Picture: Overpopulation (and Ways of Dealing with It)

*There is a passion for hunting something deeply implanted in the human breast.*
—Charles Dickens, *Oliver Twist*

*It is very strange, and very melancholy, that the paucity of human pleasures should persuade us ever to call hunting one of them.*
—Samuel Johnson

Not to overstate the obvious, but there are a *lot* of deer in the United States. As I said in chapter 1, estimates range from fifteen to twenty-five million. That's a lot of anything, let alone deer. The recent increase in population stems from many factors, including increasing suburbanization (more people mean more plantings for deer to eat), consecutive years of milder winters, a decline in the predator population (haven't seen many wolves around, have you?), a decline in the number of hunters, and other factors that are more difficult to categorize.

Many years ago deer were hunted so intensively that their population was greatly depleted. After prolonged efforts to restore their numbers, the deer population has not only rebounded but grown so large that it has become a tremendous problem. You can't say the same thing about other species, such as bald eagles or buffalo.

And it's not something that is a bother for just gardening enthusiasts, either. There are other, more serious consequences that affect us all. Overgrazing by deer stresses their entire habitat (one that they share with a great many other creatures). Overpopulation spreads disease. And once crowded out of their woodland environment, young males increasingly wander across roadways, causing car accidents, or invade homes and gardens, leaving havoc in their wake.

It's easy to get people to agree that something needs to be done. What's hard—maybe impossible—is to get them to agree on *what* that something is.

Basically, there are four choices: relocation (move them out), contraception (sterilize them), predation (get other animals to kill them), or elimination (just shoot them).

Each solution has its pluses and minuses; each has its advocates and foes. And boy, do they not get along! Before a community, city, or national park can make a rational choice as to which technique will work best for them, there are several factors to be considered: cost efficiency, long-term effectiveness, ecological balance, and social acceptability.

## Relocation

Seemingly the simplest solution to the problem of deer overpopulation in a community is to relocate the animals to another area, which some places have tried. In order to get twenty-six deer per square mile in rural areas (less in cities)—the optimal number according to many state wildlife officials—deer are rounded up and moved elsewhere.

The first decision to trap and remove deer to control the community's deer population in Missouri happened in 1998 in Town and Country. Residents had previously used sharpshooters to curb the out-of-control deer populations. The proposal called for seventy-five does to be trapped and released in public areas in the state designated by biologists. Seventy-two percent of surveyed residents supported the use of nonlethal methods to control the burgeoning deer population. A group of residents raised fifty thousand dollars in 1998 to move the deer, but the total cost of relocating was estimated to be a hundred thousand dollars. The relocation was successful, but it is impossible to know if it will help reduce the deer population over the long term.

Relocating deer is expensive and not as easy as relocating, say, squirrels. You can trick squirrels into a little trap by using peanut butter as a lure. Then, with gloves donned, it's easy to transport the critters. But try moving a deer in your Chevy Suburban.

Lakeway, Texas, has been relocating deer on and off since 1986. In 1997 alone the community relocated 450 deer. Although the trapping is free to Lakeway residents, ranchers pay the trappers a fee for delivering the deer—for increased hunting opportunities, as well as scenic beauty. State law requires that ranches that receive relocated deer have enough habitat to feed them and allow alternative population controls, such as hunting.

However, roughly 40 percent of those deer didn't survive the move; the trauma of leaving their home territory was too great. (Children going away to summer camp, fortunately, do not suffer the same trauma.)

Frequently deer experience what is called "capture myopathy" in which a trapped deer suffers shock upon being captured and moved and never recovers. The animals can go into shock from being in the proximity of handlers for as little as eight minutes. One out of every ten deer caught and later released will die of capture myopathy. Within three weeks of being released, deer develop the condition that paralyzes their back legs, leaving them to starve to death or be devoured by wild dogs or coyotes. The *Minneapolis Star Tribune* estimates that as many as 80 percent of deer captured for relocation die, either in transit or afterward.

Advocates who support relocating deer say that less stressful handling techniques will reduce the death rate. Instead of using the explosive rocket netting that fires a rocket-propelled net over the deer, for instance, they support using drop nets. Less shock to the deer means less of a chance that they will die.

Another issue is how deer will survive if they are moved from their matriarchal groups and into new environments. A deer's social structure, which usually consists of a doe and one to two fawns, can be harmed when deer are rounded up for relocation and separated from each other. In the wild deer are born, grow up, live, and die in a thirty-mile radius, so obviously a new territory might require too much of an adjustment.

And who wants the deer? There are too many deer in too many areas already—far more than the areas can support. Fifty years ago relocation might have been an effective solution, but today deer are everywhere. Park superintendents, wildlife officials, refuge managers, and landowners don't want more deer than they have already. This proposal, therefore, is perhaps not the best solution to solving one area's problem; it merely transfers the problem to another place. It's a lot like stuffing your dirty laundry under the bed—you don't think about all the smelly clothes because they are hidden from your sight, but the pile continues to get bigger and bigger.

Since wildlife experts say that trapping and relocating deer just isn't feasible—in addition to the high deer mortality rates, the procedure can cost from one hundred to eight hundred dollars per deer, and given the sheer number of animals, it's too time consuming to be a practical solution to the problem of overpopulation.

## Contraception

If we could slow down or even prevent reproduction, we could reduce the deer population. Although nature plays a large role in population control—harsh weather conditions and scarce vegetation limit the number of deer that survive—humans have been considering and implementing ways to speed up nature's cruel work. Advocates for birth control have adopted the slogan "Curb the herd."

Contraceptives are usually delivered through a blow-dart injection or by standard injection after the doe has been tranquilized. If the tranquilizer-first method is used, does must be tagged for further shots. Booster shots need to be administered a few weeks later and then once a year after that. There is currently no single-dose injection system, but efforts to develop one are continuing.

The contraceptive drug is made from a protein from pigs that in swine causes the sperm to attach to the egg, but when injected into a female from another species (such as a doe) it has the opposite effect: It triggers the doe's instinctive response against a foreign substance. She then generates antibodies that attack her own sperm-attachment proteins. The buck's sperm cannot bind to the egg and, therefore, conception cannot occur. This vaccine presents no danger to humans who eat venison.

Deer contraception received the highest rating in a survey conducted in late 1997 by the *Buffalo News* on various deer-control measures. On a scale of 1 (poor) to 5 (very good), the results were:

Contraception: 3.24
Fencing: 3
Bait-and-shoot: 2.99
Driver education: 2.82
Capture and release out of town: 2.74
Reduction of speed limits: 2.59
Hunting in selected areas: 2.45
No action: 2.22.

The *New York Times* reports that at Fire Island National Seashore, the National Park Service and the Humane Society have used birth control for deer since 1993. In three years pregnancies have gone down by 85 to 90 percent. Fire Island may present an exceptional case, however, because it is a contained environment where deer are easy to monitor. In mainland wooded areas the effi-

than 0.25 percent of livestock is available to the wolves. Ranchers are allowed to kill wolves that harm their livestock on private land. Advocates who promote the reintroduction of these predators say that they do not pose a significant threat to humans or pets.

In 1995 fifteen wolves were released into the Idaho backcountry, where abundant deer herds roam. The U.S. Fish and Wildlife Service reported signs that the wolves had killed deer in the area and have been able to live off the land.

In places like Shelter Island, New York, that have no natural predators, the deer population has gotten out of control. But it isn't practical for such communities to bring in wolves in order to reduce the number of deer.

Reintroducing the deer's natural predators may help curb the overpopulation, then, but natural predation is only practical in remote wilderness areas, such as national parks. It is not an option in rural areas—and certainly not in suburban areas, where the deer population has reached a near-crisis level.

## Elimination: A-Hunting We Will Go

*Stuffed dear heads on walls are bad enough, but it's worse when they are wearing dark glasses and have streamers and ornaments in their antlers because then you know they were enjoying themselves at a party when they were shot.*
—Ellen Degeneres

*The fascination of shooting as a sport depends almost wholly on whether you are at the right or wrong end of the gun.*
—P. G. Wodehouse

To start with, here are some facts about deer hunting:

- The hunting industry earns an estimated eleven billion dollars a year.
- There are approximately thirteen million licensed deer hunters in the United States.

- In 1995 hunters killed six million white-tailed deer, out of a total population of twenty-five million.
- Hunters most often use a rifle (or another type of gun) or a crossbow.
- Biologists say that stabilizing a population requires that 40 percent of females be killed each year.
- Hunting begins in late fall and lasts through early spring.

Deer hunting is both big business and popular. Regardless of your opinions about hunting, it is important to a lot of people and companies.

Deer hunting has been a sport and a means of putting meat on the table for hundreds of years. It's a popular pastime that will doubtless remain with us for generations to come. (In an effort to increase the number of hunters in the state, for instance, Virginia's game agency proposed new hunting rules in March 1999 to encourage more children and adolescents to hunt.) In its defense, many hunters have focused the debate on the strong evidence that their sport does indeed significantly check deer overpopulation. To cite just one example, the pro-hunting book by William J. McShea called *The Science of Overabundance* says that hunting is an efficient way to rid an area of too many deer. Furthermore, in certain rural areas in the United States deer hunting makes a vital contribution to the family table. (For more on venison as a food source, see chapter 7.)

Sportswriter Grantland Rice in response to hunting partners who called him a "sissy" when he failed to pull the trigger on a white-tailed deer: "It's not that, exactly. It's just that I never shoot a deer until he pulls a knife on me first."

Killing an animal is the goal of hunting, of course, but hunters say it is the experience itself that brings the hunter and animal together. They argue that we were born to hunt, just as deer were born to be the prey of hunters—it is the natural order of life. With-

nesota hunting accidents involved deer hunters—far more than firearm accidents affecting hunters of any other species, such as game birds or bears. Occasionally, these accidents have fatal results, such as the eight-year-old boy who was hunting with his father in Orange County, California, and was killed when his father fell; the man's rifle accidentally hit a rock and fired. In 1996 a man who was deer hunting near his home in New York State was fatally shot by another hunter who was carrying an illegal rifle and a forged hunting license.

Accidents happen because too many hunters who go out after deer or other game are careless, incompetent, or both. Homeowners complain that hunters routinely:

- ignore NO HUNTING postings
- trespass on private property
- violate state or local regulations forbidding gunfire within a set number of feet or yards of a house, barn, or garden shed
- fail to practice basic gun safety rules
- receive "slap on the wrist" fines when caught breaking the law
- hunt while drunk
- are just plain lousy shots

Hunting advocates counter that none of these problems in and of itself makes a case against hunting. These are all matters that can be remedied by better enforcement of existing regulations, by handing out stricter penalties to those who break them, by increased funding for state wildlife and game agencies, by the hiring of more game wardens, and perhaps by requiring training courses for hunters before they can be licensed. In other words, work to make hunters better at what they do—don't stop them from doing it.

The National Rifle Association recommends the following basic firearm safety rules:

- Point muzzle away from yourself and others.
- Keep safety in ON position until ready to shoot.
- Never rest finger on trigger.

- Unload firearms when not in use.
- Never step over an obstacle with a loaded firearm.
- Always treat a firearm as if it's loaded.

According to a Reuters story in 1996, in an effort to catch hunters who use illegal weapons, shoot female deer, or hunt their prey without authorization, West Palm Beach, Florida, game wardens began using two new battery-operated decoy animals called Robo-Deer and Robo-Bear, collaring fifty-seven hunters that year. The full-sized models, created by a taxidermist, use the head and hide from real animals and even nod their heads and wiggle their tails. Game officers reported some strange events resulting from the use of the decoys, such as two poachers attacking each other during an argument in which they criticized each other for being lousy shots and missing the Robo-Deer that they thought was real.

Sometimes hunting clearly goes too far. Ingrid Jakens, a friend, remembers a deer-hunting incident in Northampton, Massachusetts, a few years ago:

> We have a park in the center of the city, a place to walk your dog, have picnics, let your children play. They have this small zoo there—only a few birds (mostly owls), maybe a sheep, and three deer. Definitely whitetails, enclosed by a high chain-link fence. One morning, the park maintenance staff found one deer dead and another wounded by arrows. Someone had gone in and shot the enclosed deer, but in the middle of the park—that's sick. If someone enjoys hunting, that's fine. But to kill a defenseless animal that can't even run away—where's the sport in that?

Those who feel endangered by the seemingly reckless hunting that goes on around them will probably want to see the system tightened. They may want to press for still-tougher measures, such as requiring hunters to pass a marksmanship test before they can be licensed. After all, we make drivers demonstrate that they can actually handle a car on a road test before we license *them*—why

"urban deer zones" were declared in which controlled hunting
was allowed in areas around Cincinnati, Cleveland, Akron,
Youngstown, and Columbus, Ohio, with the result that an estimated forty thousand deer were killed, according to *USA Today.*
Outside Chicago, in the cities of Lake Forest, Glenco, Northfield,
and Highland Park, state-certified marksmen have been hired to
shoot deer in wooded areas.

Montgomery County in Maryland, just outside the District of
Columbia, has been trying unsuccessfully for five years to control
the deer population. The county began authorizing deer hunts in
1996. Efforts such as repellents, electronic fences, and ultrasonic
devices have all failed. (The *Washington Post* described how deer
outwitted the humans by waiting outside the electronic gates until
a car pulled through, then following it inside.) The Montgomery
County herd, said to number between fifteen and twenty deer, has
been eating its way through the fifty-acre Wheaton Park, one of the
most popular parks in the region with more than three hundred
thousand visitors each year. Officials report that in a two-year
period 18,400 tulips, 770 pansies, and 457 rose plants were eaten
by deer. Officials took the controversial step and sharpshooters
began eliminating the deer in February 1999.

But where culling goes on—even in those places where it has
been shown effective (or rather, *especially* in those places)—opposition follows. Police sharpshooters in 1996 had killed 858 deer in
Gettysburg National Military Park when they were forced to stop by
a suit filed against the practice by animal rights activists in federal
court.

Hunting opponents are a busy bunch—and they're wired, too.
Check out the Animal Rights Resource Web site at aars.envirolink.org,
which is run by the EnviroLink Network and the International Fund
for Animal Welfare. In a recent campaign to end the fourth annual
"Bait and Shoot" program in the Irondequoit/Rochester, New York,
area, during which 455 deer have been killed, the organization
called for an economic boycott of Irondequoit businesses as well as

Rochester's convention, tourist, cultural, and sporting events until the program is canceled.

In Montgomery County, Maryland, demonstrators gathered in 1996 at Seneca Creek State Park and told the *Washington Post* that they wanted a "blessing of the deer"—complete with a minister sprinkling holy water—to protest planned hunts in the county.

The 250 retired nuns at Mount Saint Francis Center, in Dubuque, Iowa, however, feel differently about hunting, according to the *Des Moines Register.* Although the nuns "want to live at peace with the creatures of God," administrator Connie Tjarks says, "they in turn have to give us some respect, too." Deer have been terrorizing the ten acres that the nuns farm for food (including grapes, beets, squash, tomatoes, and cabbage), and to combat the problem the sisters have asked the Dubuque City Council for permission to allow hunters on their 120 acres of land to decrease deer damage.

It's not all that difficult to see why culling, as opposed to the more usual recreational hunting, should stir up such intense opposition. Many of us, even if we don't hunt ourselves, feel a certain (grudging, at times) admiration for those who do. We can understand the primordial need to go out into the wilderness, be one with nature, stalk silently, and take down a large creature through your pluck and skill—to be the one who quite literally puts food on the table. Scientific culling—especially of animals herded into enclosures—seems ruthlessly efficient and utterly heartless.

When it comes to the battle for public sympathy, the animal activists with their posters of wide-eyed does doomed to a bloody death win hands-down every time. When communities decide against controlled hunting, voters may congratulate themselves for making the humane choice, without giving any further thought to what the future will be like for those deer that must go on living in a region burdened by more animals than the land can support. How will they feel when the herd starves all winter? What's it like to

# Hunters in America *(Continued)*

| State | Estimated Population (as of December 1996) | Licensed Hunters | Hunters as a Percentage of Population |
|---|---|---|---|
| Arkansas | 2,510,000 | 332,000 | 13.2 |
| California | 31,878,000 | 505,000 | 3.6 |
| Colorado | 3,823,000 | 237,000 | 6.2 |
| Connecticut | 3,274,000 | 46,000 | 1.4 |
| Delaware | 725,000 | 27,000 | 3.7 |
| Florida | 14,400,000 | 170,000 | 1.2 |
| Georgia | 7,353,000 | 353,000 | 4.8 |
| Hawaii | 1,184,000 | 23,000 | 1.9 |
| Idaho | 1,189,000 | 183,000 | 15.3 |
| Illinois | 11,847,000 | 397,000 | 3.4 |
| Indiana | 5,841,000 | 338,000 | 5.8 |
| Iowa | 2,852,000 | 297,000 | 10.4 |
| Kansas | 2,572,000 | 212,000 | 8.2 |
| Kentucky | 3,884,000 | 348,000 | 9.0 |
| Louisiana | 4,351,000 | 323,000 | 7.4 |
| Maine | 1,243,000 | 148,000 | 12.0 |
| Maryland | 5,072,000 | 110,000 | 2.2 |
| Massachusetts | 6,092,000 | 80,000 | 1.3 |
| Michigan | 9,594,000 | 865,000 | 9.0 |
| Minnesota | 4,658,000 | 544,000 | 11.7 |
| Mississippi | 2,716,000 | 291,000 | 10.7 |
| Missouri | 5,359,000 | 471,000 | 8.8 |
| Montana | 879,000 | 141,000 | 16.0 |
| Nebraska | 1,652,000 | 131,000 | 8.0 |

| State | Estimated Population (as of December 1996) | Licensed Hunters | Hunters as a Percentage of Population |
|---|---|---|---|
| Nevada | 1,603,000 | 46,000 | 2.9 |
| New Hampshire | 1,162,000 | 66,000 | 5.7 |
| New Jersey | 7,988,000 | 84,000 | 1.0 |
| New Mexico | 1,713,000 | 88,000 | 5.1 |
| New York | 18,185,000 | 608,000 | 3.3 |
| North Carolina | 7,323,000 | 313,000 | 4.3 |
| North Dakota | 644,000 | 77,000 | 12.0 |
| Ohio | 11,173,000 | 443,000 | 4.0 |
| Oklahoma | 3,301,000 | 284,000 | 8.7 |
| Oregon | 3,204,000 | 272,000 | 8.5 |
| Pennsylvania | 12,056,000 | 752,000 | 6.2 |
| Rhode Island | 990,000 | 19,000 | 1.9 |
| South Carolina | 3,699,000 | 238,000 | 6.4 |
| South Dakota | 732,000 | 109,000 | 15.0 |
| Tennessee | 5,320,000 | 362,000 | 6.8 |
| Texas | 19,128,000 | 829,000 | 4.3 |
| Utah | 2,000,000 | 113,000 | 6.0 |
| Vermont | 589,000 | 68,000 | 11.6 |
| Virginia | 6,675,000 | 363,000 | 5.4 |
| Washington | 5,553,000 | 256,000 | 4.6 |
| West Virginia | 1,823,000 | 252,000 | 13.9 |
| Wisconsin | 5,160,000 | 586,000 | 11.4 |
| Wyoming | 481,000 | 67,000 | 14.0 |
| **Total** | **265,284,000** | **13,330,000** | **5.0** |

The deer debate has provided our language with new words and phrases. These are some of my favorites, which represent both sides of the deer debate; I call them New Slogans for the New Millennium:

- CURB THE HERD
- A FAWN OR A LAWN?
- DOE, A DEER, A FEMALE NUISANCE
- STAG PARTY—DOES GET IN FREE
- DEERLY BELOVED
- BLEAT, BELLOW, AND BAWL: THE SOUNDS OF THE WILD
- NO ARROW IN THE MARROW
- TED NUGENT SAYS, "HAVE YOU KILLED A DEER TODAY?"
- SMILE, IT'S CHEAPER THAN A BULLET
- STUNT THE HUNT
- DEER NEED LOVE, TOO

# Learning to Love Them

*As much as it is important to provide the most humane solutions to problems between wildlife and people, we must also take action to preserve the habitat bases on which so much wildlife depends.*

> —Paul G. Irwin, president of the Humane Society of the United States

The emergency room staff at Hadassah University Hospital in Jerusalem reacted quickly when someone rushed in holding a swaddled three-year-old. Two orthopedic surgeons left a medical conference to attend to the young patient, later given the name Hadassah. During an hour-long operation at nearby Hebrew University Hadassah Medical School, another pair of doctors worked feverishly to fix the smashed tibia in one of the patient's legs and sew up a giant cut.

After the surgery doctors brought young Hadassah to the Tisch Family Zoological Garden, where she joined the zoo's twelve other deer. Yes, that's right—Hadassah is a doe. A good Samaritan saw the deer fall from a cliff in Sataf and rushed her to the nearest hospital, which just happened to be for humans. Chief veterinarian Dr. Gabi Eshkar called the doctors "physicians with a soul."

Deer have that effect on people.

In its book *Wild Neighbors: The Humane Approach to Living with Wildlife,* the Humane Society of the United States says:

> One of the best ways to address current problems, as well as to look ahead to future coexistence with deer, is to encourage understanding and a tolerance for these animals and the impacts they sometimes have on resources that humans seek to protect. This certainly is not to say that all of the damage that deer might cause has to be accepted, but only that it is inevitable that *some* will occur where deer and people share living space. . . . There are some knotty problems ahead of us in our relationship with deer that will have as much to do with the values and attitudes we hold about these animals as the demonstrable facts about their interactions with their environment. We must acknowledge that these animals will be a continuing part of our lives. Let's start by accepting and appreciating them for what they are before seeing them as a problem-causing crisis that has to be "solved."

Well, I wouldn't have put "solved" in quotation marks. I don't see anything wrong with seeking solutions to the deer problem, but I agree with the Humane Society that we've got to be realistic about how well any one solution will work. In the long run—unless you give up and move to midtown Manhattan—you'll still be there and so will the deer, so you'd better get used to each other.

It might help to remember that *live* is only one letter away from *love.* When you're finished mulling over that one, move on to the next—*love* doesn't mean allowing someone to freeload off you. Love means not being a doormat. We're talking "tough love." Is that better?

When it comes to garden-destroying deer, what do tough-love advocates do? As with relationships among people, there is an infinite variety of expressions of feelings. Some feed the deer in a protected habitat to draw them away from gardens. Others play games with them—video games, virtual deer hunts that allow a bloodless expression of their disapproval of bad deer behavior toward humans. Others really, really love deer—on the dinner plate; they cook and eat venison.

I'm not here to tell you to do any of these things—just to point out that you can.

People express their feelings about deer in myriad ways: Bill and Shelley Collins like the idea of deer in gardens—a sculpture of a deer, that is. The couple started and run Ironman Recycled Metal Sculpture in California. They create animal sculptures out of old scrap metal and tools, including bicycle frames and parts, wheels, scissors, and garden rakes. What started as a hobby has earned the couple some art prizes, "a few pennies," and an article in the *Ventura County Star* in 1997. Although they declined an offer from an overseas company for their work, the couple is thinking about opening a store to sell the art exclusively.

## This Land Was Made for You and Me

A deer with a chain of gold, if she escapes, will run off to the forest to eat grass.

—Malay proverb

Wildlife specialists have recognized that the current deer population has reached the limit at which the natural environment can sustain them. Thus, many people have created parks and preserves so that deer may continue to have a comfortable—and protected—living space.

The only way to be sure deer won't invade our turf is to work to keep and protect *their* turf, the wilderness where no humans are allowed to settle. The best natural deer habitat would be one of several thousand pristine wooded acres, far from any human development, where natural predators could perform the population-control job nature designed them for. Work to expand state or federal parkland in your area and you will be doing a good turn not just for deer but for future generations of people as well.

But what about those who live in smaller, more urbanized regions where wilderness areas are limited and cannot be expanded? Private refuges or conservancies may be the solution. Nonprofit wildlife organizations set aside land for deer and other woodland animals and allow visitors to camp, hike, enjoy the beauty of nature, watch the deer eat—and then leave them in peace. Nature conservancies may do fund-raising to purchase land, may receive land through a will or a philanthropic gesture by an individual, or may arrange with governments or private developers to swap developable land for the conservation area to be preserved.

Another solution is a joint venture between home contractors and wildlife specialists. Rick Hettinger, a conservation officer in Kentucky, suggested this solution to the *Courier-Journal:* "[Deer] need a travel corridor in subdivisions so they can run between large farms. When developers go in and cut all the trees down and eliminate travel corridors, they create problems. The deer become isolated and concentrated." Citizens can work to change zoning and subdivision regulations to require that such wooded enclaves be preserved. The alternative is deer that, after losing their natural homes to the bulldozer, will begin to eye yours with jealousy. They may not be able to get past your front door, but you can bet they'll have no trouble getting by your back garden gate.

Judy Scott, a gardener in Ithaca, New York, learned to love her garden pests. She had this to say to an Internet newsgroup:

> I just think gardening for the birds greatly increases the pleasure of being outdoors. Previous owners landscaped

so *everything* perennial is either food or shelter for birds—from highbush cranberry and other viburnums, the much-maligned thistle, blueberries, many cherry varieties, and firethorns to thirty-five-foot spruce and other assorted evergreens. The deer feast on the apples on the periphery and leave all else alone. As a result of this strategic planning, we have no mosquitoes, and have never had a bug infestation of any kind. The birdsong is glorious and the birds [are] fun to watch with the icing on the cake—about twenty hummingbirds.

Some states are making an effort to let taxpayers use their dollars to make a statement. Animal-friendly states like Massachusetts and Florida give those registering their cars at the Department of Motor Vehicles a "save the planet" option. For a small donation people can choose a specialty license plate; this provides animal rescue and preservation groups with not only extra funds but free advertising as well. For example, in 1997 Georgia opted to GIVE WILDLIFE A CHANCE, selling 528,000 automobile tags with this expression at fourteen dollars each and raising about seven million dollars for wildlife conservation projects and habitat preservation. Georgia also collected $258,000 in state income tax donations to help preserve wildlife in 1996.

For some people, like Walter Pavluk, nothing but acceptance works. As owner of Uncle Wally's Tree Farm, he watched deer damage roughly 150 evergreen trees each year. A couple of years ago Walter and Savina, his wife, decided to turn a bad situation into a happy one. The couple now sell the "deer tree," a miniature Christmas tree perfect for tabletops and apartments. The Pavluks cut the tops off deer-eaten trees to make the little versions of the mainstay of their farm, chop up the rest of the trees for firewood, and leave the branches out for people to use for garlands and wreaths. "Nothing goes to waste," Walter Pavluk told *The Plain Dealer* in 1997.

# When You Can't Beat 'Em, Eat 'Em

## Venison as a Food Source

The market for venison is growing 30 percent each year in the United States. And the U.S. supply provides only 25 to 30 percent of that ever-growing demand, according to the North American Deer Farmer's Association.

Venison has been a food source worldwide for as long as deer have roamed the earth. Besides the sport of hunting, the reward—venison—is valuable.

Venison is so important a cash crop that shortly after the kill, many hunters practice "gralloching," a process by which they remove the organs and intestines of the animal to keep the meat pure.

I got the following recipes from SOAR—the Searchable Online Archive of Recipes, an excellent Web site (soar.Berkeley. edu/recipes/) filled with more than 46,675 different things to cook.

### Curried Venison

*Yield: 6 servings*

| | |
|---|---|
| 1 ½ medium onions, minced | ¼ teaspoon Tabasco sauce |
| 3 celery stalks, chopped | ½ teaspoon Worcestershire sauce |
| 2 apples, minced | 2 cups stock or bouillon |
| ¼ cup salad oil or shortening | 2 tablespoons flour |
| 2 teaspoons curry powder | 2 pounds cooked elk or deer, cubed |
| 1 teaspoon salt | 1 cup cream or canned milk |
| ⅛ teaspoon pepper | 1 egg yolk, well beaten |
| ¼ teaspoon ginger | 3 cups boiled rice |

Sauté the onions, celery, and apples in oil until slightly brown. Stir in the curry powder and simmer for 5 minutes. Add remaining seasonings and stock and cook for 20 minutes more. Stir in the flour mixed with stock and cook for 5 minutes, stirring until thickened. Remove from the heat and allow to stand for 1 hour. Reheat and add the cooked meat, cream or milk, and egg yolk just before serving. Heat to the boiling point, stirring constantly. Serve over rice.

*Source: Agriculture Extension Service of the University of Tennessee Institute of Agriculture, submitted by Larry Christley on October 6, 1993.*

## Venison Shortcake

*Yield: 6 servings*

| | |
|---|---|
| *Butter or margarine* | *2 tablespoons tomato catsup* |
| *1 slice bacon, diced* | Shortcake |
| *¼ cup sliced onion* | *2 cups flour* |
| *1 pound ground elk or deer* | *2 teaspoons baking powder* |
| *½ teaspoon salt* | *½ teaspoon salt* |
| *¼ teaspoon pepper* | *¼ cup shortening* |
| *2 tablespoons flour* | *⅔ cup milk* |
| *1 ¼ cups water* | *Butter or margarine* |
| *½ teaspoon prepared mustard* | |

In a bit of melted butter or margarine, sauté the bacon and onion until slightly browned. Add the meat, salt, and pepper and cook until browned. Add the flour and blend, then add the water, mustard, and catsup. Bring to a brisk boil, stirring constantly.

To make the shortcake, preheat the oven to 425°F. Sift the flour, baking powder, and salt together twice. Cut in the shortening. Add the milk gradually, mixing to soft dough. Knead slightly. Roll ¼ inch thick and cut with a floured 3-inch biscuit cutter. Place half the biscuits on baking sheets, brush with melted butter, and place the remaining biscuits on top. Bake for 12 to 15 minutes.

To serve, split each shortcake and pile the meat mixture between the halves.

*Source: Agricultural Extension Service of the University of Tennessee Institute of Agriculture, submitted by Larry Christley on October 6, 1993.*

If you're looking for something a bit more traditional, try this recipe for chili from *The Great Chili Book* by Bill Bridges.

LBJ Pedernales River Chili

*Yield: 8 servings*

*4 pounds bite-sized or chili-ground venison,* or *well-ground chuck*
*1 large onion, chopped*
*2 garlic cloves, minced*
*1 teaspoon ground oregano*
*1 teaspoon cumin seeds*
*2 cups hot water*
*6 teaspoons chili powder, or to taste*
*1 ½ cups canned whole tomatoes*
*2–6 generous dashes of liquid hot pepper sauce*
*Salt to taste*

Place the meat, onion, and garlic in large, heavy skillet or Dutch oven. Cook until light colored. Add the oregano, cumin, water, chili powder, tomatoes, hot pepper sauce, and salt. Bring to a boil, lower the heat, and simmer for about 1 hour. Skim off the fat during cooking.

## How to Tell if You're Eating a Good Deer

Here is some advice from deer hunters on how to cook a good deer:

- Keep the deer from tasting too gamey by properly field-dressing.*
- Cool the carcass quickly.
- Don't cook a deer killed during its rutting (or breeding) season.
- Don't cook a visibly old deer.
- Don't cook deer that have been subsiding on a diet of pungent or aromatic plants.

*Hunters "field-dress" a deer after it has been killed. The skin is cut open and the organs (with the exception of the lungs, liver, and heart) are removed.*

- Tenderize gamey meat with marinades and strong-flavored ingredients like chili powder, curry, garlic, and gingerroot.
- Never overcook; since deer have less fat than most domesticated animals, less cooking time is needed per pound.

"Garçon, I'd like to start with deer hearts as an appetizer. Then venison as an entrée, and a little mousse for dessert." Yes, it's true: Some hunters keep and pickle deer hearts to serve as an appetizer.

## Venison for Charity

Charities have begun to take advantage of venison as a food source. In Maryland alone close to twenty-five tons of venison have been donated by hunters and farmers, which equals about 160,000 meals. An organization called Farmers and Hunters Feed the Hungry (FHFH), with the financial support of churches, businesses, and other organizations, processes and distributes venison and other meats to the hungry of Maryland. FHFH was formed in 1997 as a pilot program in Washington County, using Virginia's successful Hunters for the Hungry program as a model. FHFH is the only program of this kind endorsed by the Maryland Department of Natural Resources.

Farmers and hunters deliver the harvested surplus deer to participating meat processors in each county. Donations from churches, clubs, businesses, and individuals cover the costs of processing, packaging, and freezing the meat, which is then picked up from the participating meat processors and delivered free of cost to food banks located throughout the state. Visit FHFH's Web site at www.hfth.org

Westchester County, New York, has instituted a policy of giving hunters a bonus for killing females in order to cut down on future deer reproduction. The Westchester County Bowhunters Association, with two hundred members, sponsors a program called

"Hunters Helping the Hungry," which donated two thousand pounds of venison to the needy from 1994 to 1997.

Project Venison, established by the Buckmasters American Deer Foundation, provides more than half a million meals to the needy and works to raise money through events like golf tournaments. The group served roughly three thousand individuals in 1991, its first year of operation. This is the donation process: A hunter brings the dead deer to an approved deer processor. The processor grinds the meat into something resembling hamburger and donates it to the local Second Harvest food bank affiliate. Buckmasters then pays the deer processor from money raised in raffles, auctions, and events.

Although not a solution to the deer problem that vegetarians would endorse, these meat-donation programs are winning increasing support from many different sections of society: deer-plagued homeowners, hunters, wildlife managers, and, of course, the needy themselves who get a gourmet dinner of venison for a change, instead of the usual bland soup-kitchen fare.

## Deer Farming

After a whole chapter on the problems of deer overpopulation, you may wonder about what I'm going to report next: There are people in the business of breeding *more* deer. Right now you're thinking, "Why are these yo-yos breeding more deer? I'd be happy to give them all of mine."

Actually, this is nothing new—deer ranching and farming began in Asia and Europe more than four thousand years ago. (Its roots can be traced even farther back to the Mesolithic era, 40,000 to 10,000 B.C.) However, only in the past two decades have deer breeding and farming come to the United States.

One modern deer operation is Bonnie Brae Farms in Plymouth, New Hampshire (the gateway to the White Mountains and Lakes Region), which began in 1994 with twenty-seven red deer.

Located on two hundred acres of old farmland, Bonnie Brae has been part of the family for three generations. Its goal is to increase sales of deer meat and raise public awareness of the health benefits of venison. Other deer farms are located in Maine and Vermont.

Venison World out of Eden, Texas, was founded in August 1992 by seven ranchers who raise exotic deer and antelope on thousands of grassy, live oak, and agarita-covered acres. From their Web site:

> These ranchers turned to deer ranching in order to propagate adaptable popular and threatened species, to make better ecological and economical use of their range, to improve aesthetics, to enhance recreation, and to promote a new Texas industry.

See chapter 9, Resources, for contact information.

## Games People Play

Would-be deer hunters have turned out to be a huge market for manufacturers of games for personal computers. (The game Deer Hunter II even beat Microsoft Windows 98 Upgrade in 1998 as the top-selling CD-ROM.) Are the folks who purchase these games using them for off-season training? Or are people who wouldn't necessarily go out and kill a deer using them as a harmless substitute for the act? What do the huge sales of Deer Hunter and Deer Avenger tell us about society today?

Well, I'll let the pundits answer that one. The sales figures themselves demonstrate that people are definitely still connected to deer—at least on some level.

### Deer Hunter

You can find out why this game is so much fun by downloading a free copy of Deer Hunter at Ziff-Davis's Web site, www.zdnet.com, and elsewhere on the Internet. Deer Hunter is also available from GT Interactive for about twenty dollars.

The original Deer Hunter game, which debuted in October 1997, has sold more than one and a half million copies and was the top-selling personal computer game for January, February, and March 1998. Only seven months after its introduction the game had earned a total of $8.1 million.

Deer Hunter was the first game to simulate actual hunting, including the sounds of the soft whistling wind, distant chirping birds, and rustle of grassy fields. Rattling antlers and mating calls attract the deer, but you must also watch for wind direction in order to keep your scent out of the animal's path. Earlier hunting games often resembled video-arcade shoot-'em-ups more than real hunting. Part of Deer Hunter's attraction is its realistic features.

Choose either a rifle, shotgun, or bow as your weapon, and make use of a treestand, cover scent, and attractant scent.

Deer Hunter II, available from WizardWorks (www.wizard works.com), was released in October 1998 and rocketed to number one on the *PC* charts of interactive games.

## Deer Avenger

In a unique twist on the traditional hunter-versus-deer scenario, Deer Avenger, an interactive CD-ROM game available from Simon & Schuster Interactive and Cendant Software ($19.95), lets you be the deer in a parody of the successful Deer Hunter game.

Assume the identity of an animated cartoon buck named Bambo and begin to stalk your prey—orange-clad hunters, rednecks, and tree-huggers who leave toilet paper, nudie magazines, beer cans, tree carvings, and markings in the snow such as KLEM WUZ HEAR. Bambo utters such cries as, "Help, I'm naked and I have a pizza!" and "Free beer here" to attract his prey. His mating calls range from "Sweet! *Baywatch* is on!" to "Ted Nugent's here!" to "Viagra!"

Neither a glorification nor condemnation of hunting, this humorous game was written by *Late Night with Conan O'Brien* staff writer Brian McCann, who has been quoted as saying, "[Deer Hunter] was just screaming to be made fun of. . . . It's just so boring; it's even less interesting than a board game."

Safari Club International, which has thirty-two thousand members in forty countries, wants Deer Avenger recalled. Alfred Donau, president of the organization, says hunters are being misrepresented as "stupid, drunken womanizers." Okay, but what about the deer? Who's suing on behalf of poor maligned Bambo?

# When All Else Fails: Dastardly Deeds That Can Deter Deer

**8**

*It is only by being obstinate that anything is got, or done.*
—Rumer Godden

All right, you've gotten this far on your quest to eradicate deer from your life—well, at least your backyard. You've tried noisemakers, you've erected a monstrosity of a fence, the sprinklers are running, and still each morning and evening you find evidence that your favorite four-legged friends have found yet another safe entry point to access and enjoy your property and its plantings.

So you've learned that although deer may not be rocket scientists, they are very, very good at doing what deer are supposed to do—find food. In fact, they really do very little else all day (except in mating season) beyond that. In other words, that's their job. To do it well, they need to overcome all obstacles that we humans put in their path—that is to say, outwit our cleverest schemes to outwit *them*.

Then we strike back with still harsher, more devilish countermeasures, and the war escalates—us against Bambi and the four-legged troops. Neither side will ever surrender. Why should we? It would mean losing our greenery, our privacy, our pride of place, our "home sweet home" feeling that has inspired everyone from Dorothy in Oz on down to this present day. And the deer? Well, *they* will never give up either. Because nature designed them that way.

Where does that leave us? It leaves us having to keep coming up with new ideas, to keep trying things out. We take what works, run with it for a while, and then, after the deer adapt (as evolution decrees they will, in time), move on to the next strategies.

To make sure you have plenty of tricks up your sleeve—and more to fall back on when the first, second, and third rounds give way to deer persistence—I offer you the following list of 101 cunning stratagems to outwit deer.

Let's all give a cheer,
to the dear little deer.
Let's toast their persistence and their grit.

Let's give them their due;
and then get a clue,
that Bambi we can *always* outwit!

—Sharon Davis

1. Build fences. As big and expansive as you can erect.
2. Keep a mean dog. Either unleashed—if the local laws permit—or on a long-enough leash to give the dog access to your garden.
3. Surround the plants that deer like (such as hickory and English ivy) with plants that deer don't like (zinnias, spireas, and willows).
4. Find a nearby university biology department that's interested in experimenting with deer birth control; schools like practical scientific projects with potentially practical results.

5. Erect motion sensor lights.
6. Invite your neighbors' dogs to relieve themselves in your garden.
7. Install motion sensor sprinklers in your garden.
8. Collect human hair from local barbershops and beauty salons. Leave it in your garage for a while. Then sprinkle it on your garden. Vary the type of hair you use.
9. Encourage your community to round up all the deer and ship them somewhere far, far away. And if that doesn't work, get together at a town meeting and discuss other ways to deal with the deer overpopulation in your area that everyone feels comfortable with.
10. Experiment with products like predator urine and deer repellents. Vary the methods you use and see what works best in your area.
11. Surround your garden with things that deer hate: mothballs, human hair, the feces of their predators, and coyotes, for example.
12. Set up ultrasonic devices around your yard. The sound will keep deer away. Remember though, this will only work for a short time; deer will eventually realize that despite the horrible sound, nothing happens.
13. Hang wind chimes.
14. Cover your plants and garden in a gigantic net.
15. Try visual deterrents: flags, ribbons, anything that will flap in the wind.
16. Ring your garden with fishing wire strung two feet off the ground.
17. Use plants with strong odors in your garden. Try cedar, oregano, sage, and mallow.
18. And try to avoid planting things deer do like: pumpkins, hickory, apples, and alfalfa.
19. Be creative. Keep trying. Don't give up.
20. Offer free Gatorade to your local high school track team, but make sure that they come over right after a long run and before showering.
21. Breed and raise coyotes.

66. Attract deer to electric fencing by stringing it with peanut butter-smeared aluminum foil flags—one zap and they won't return.

67. Hang white tape ribbons at various intervals on your fence.

68. Hang bars of scented soap from sacks and distribute liberally around your garden.

69. Sprinkle cayenne pepper on deer-favored flowers.

70. Shave slivers of Irish Spring soap and sprinkle around the base of deer-attracting plants.

71. Alternate Tabasco sauce with mothballs as a deterrent.

72. Spray a mixture of clove oil, dish soap, and dormant oils on plants.

73. Splash leaves with a mixture of raw eggs and skim milk.

74. Spray castor oil on leaves.

75. Empty your vacuum-cleaner bag on your flower bed.

76. Get a hold of bear droppings and sprinkle throughout garden.

77. Next time the circus comes to town, load up on tiger and elephant manure.

78. Buy or beg coyote and/or wolf urine and leave sponges around your yard that are soaked in the stuff.

79. Hang bloodmeal in cloth sacks around your yard.

80. Enclose your garden in a giant plastic bubble. The greenhouse effect will help your plantings. Alternatively, consider an actual greenhouse. Greenhouses have a number of things going for them: They keep deer out (not to mention squirrels and other pests); keep your garden going year-round; and let you innovate in ways you never thought possible—they move you up the next level of gardening.

81. Urinate around the perimeter of your garden.

82. Set up a scarecrow near the deer's favorite plants.

83. Set up a big-screen television in your yard and play the scene from Bambi where Bambi's mother dies over and over again.

84. Tie plastic bags to your tomato stakes.

85. Install strobe lights in your backyard.

86. Tie balloons to tree limbs and let them bob in the wind.

87. Hide a motion-sensor sprinkler behind a scarecrow.
88. Place cedar benches at strategic spots in your garden.
89. Become a predator advocate—protect bears, coyotes, cougars, and wolves.
90. Consider supporting regulated, responsible deer hunting in your area.
91. Casually place copies of this book between plants.
92. Read about deer: The more you know about them, the better equipped you'll be to outwit them.
93. Move to an area that doesn't have a deer problem.
94. Protect wilderness areas!
95. Learn to love 'em. Place a feeding station in your yard— they might even ignore your garden.
96. Give your neighbors a gift certificate to the local nursery, but specify that it can be used to purchase only certain plants (which happen to be the ones that deer love).
97. Ask your city or town to build a highway through your backyard. Not only will this keep deer away, but it will shorten your commute!
98. Open your own taxidermy shop. Display your handiwork around your home.
99. Sleep outside, in a tent. Keep a watchful vigil.
100. Contact one of the many companies listed in this book that specialize in products that deter deer.
101. Use a multifaceted approach: Plant to repel deer, deter them with repellents, and fight to control deer populations in your neighborhood.

# Resources

**9**

## Gardening

*Deer Resistant Plants for Ornamental Use,* University of California Cooperative Extension Leaflet 2167, is available by calling 415-726-9059.

The Plant Lady (with deer-resistant perennial information) can be found online at www.members.tripod.com/~dlee13.

LandscapeUSA.com has a large online gardening store at www.landscapeusa.com/. It features such deer-repellent products such as predator urine and invisible black mesh barriers.

A gardening Web site at www.tenforward.com/iris/gardening.html has tips for discouraging deer browsing.

More gardening-with-deer tips can be found at www.slugsandsalal.com/techniques/deer.html. This Web site includes descriptions of fencing, ground barriers, deer-resistant plants, and repellents.